SHOOT ORGANIZATION
IN VASCULAR PLANTS

Shoot Organization in Vascular Plants

K. J. DORMER
Reader in Botany,
University of Manchester

Syracuse University Press
1972

First published 1972
by Chapman and Hall Ltd
11 New Fetter Lane, London EC4P 4EE
Published in the U.S.A. by
Syracuse University Press
Library of Congress Catalog Card Number: 70–39412
ISBN: 0–8156–5032–9

Printed in Great Britain by
Richard Clay (The Chaucer Press) Ltd
Bungay, Suffolk

Contents

Preface

In this book I have tried to set down some of the more widely applicable principles governing the growth and behaviour of the shoot system in the higher plants, and to illustrate these principles by reference to the research literature. Such an enterprise in so small a compass can only proceed upon a selective basis. My own predilections have always been for quantitative rather than purely descriptive methods of study, and for deductive inference from mature structures in preference to the direct observation of embryonic or meristematic phases. The book accordingly has a bias in these directions, and I have further thought it right to give prominence to several topics which have not been commonly discussed by the writers of other textbooks.

The study of any original investigation in the sciences appears to me to call for exactly the same powers of scholarly criticism as are ordinarily expected of a historian or biographer. The choice of a particular kind of specimen for observation or experiment must often have depended on trivial or accidental circumstances and need not always be very seriously regarded. We should be foolish, for instance, in considering the venation of leaves, to deny ourselves the benefit of cognate researches on the petal of a flower. On the other hand the contribution of an early worker will not be fairly judged if we do not allow for the inevitable confinement of his ideas within an intellectual perspective which was smaller than our own. In examining an original document it may be appropriate for us to rearrange the observations or rework the calculations. Such measures are not to be understood as implying any censure, but they may constitute belated justice, and may most interestingly reveal how narrowly an objective was missed, by how fine a margin a significant relationship escaped detection.

The execution of my design has required in places a rather extended treatment of particular examples of morphological investigation, but I hope that none will be so misguided as to see in my own imperfect text a substitute for the diligent examination of original papers. The limitations which so many students of botany impose upon themselves by their unwillingness to read German or to master simple mathematical expressions can naturally not be concealed in a work of this nature; being largely self-taught in both respects I have aimed to encourage the hesitant, rather than deter them.

Many of the phenomena presented are such as will very readily lend themselves to purely physiological enquiry, but the concern of the morphologist is not so much with the mechanism and physico-chemical basis of an effect as with the way in which it is distributed. Our subject is essentially a kind of engineering: given the elementary physiological systems (geotropism, photosynthesis, mitosis, etc.) there remains the problem of finding how these are put together to make a plant. It is to the understanding of that problem that this book aims to contribute.

K.J.D.

Chapter One

The Initiation of New Shoots

In angiosperms the expected sites for the emergence of new shoots are the leaf-axil, or, for the plumule, the apical end of the hypocotyl. The possession of a plumular shoot, though usual, even in species in which the plumule contributes nothing to the shoot system of the adult, is not absolutely universal (see pp. 21–22). The occurrence of axillary shoots is also variable, and the generalizations which can be made about them are sharply circumscribed. The axillary meristem from which buds may be produced is presumably in all cases a remnant of the apical meristem of the parent shoot. In some cases there is no such remnant and the axillary area differentiates as part of the ordinary surface of the stem. Lack of axillary meristem is more frequent in cataphylls (bud-scales, bracts, etc.) than in foliage leaves, but budless foliage leaves are by no means rare. In a number of cases there is some spatial separation between the axillary meristem patch and its axillant leaf, so that buds appear in positions which might be regarded as internodal. Extreme cases of this occur in connection with the inflorescences of Solanaceae, but although the longitudinal displacement of the branches (sometimes by a distance exceeding the length of an internode) is certainly very curious it does not involve any fundamental breakdown of the axillary relationship; each site of branch initiation continues to be uniquely and distinctively associated with one particular leaf.

It is very common for the axillary meristem to generate more than one branch apex. The buds of one leaf then form some kind of orderly pattern, almost invariably with an obvious graduation of age, as for example in a vertical series with the younger buds below (abaxial to) the older ones. The various situations which are possible have been systematically described (Sandt, 1925), and are characteristic of particular families. Experimental and genetical work has been undertaken in some cases of commercial significance (Keep, 1969).

Axillary branching is overwhelmingly predominant in angiosperms, though not among the lower vascular plants, which display other patterns of behaviour not specially considered in this book. Even in angiosperms, however, there are many examples of branch initiation from sites which are evidently not part of any axillary meristem. The ability to start the growth of a new shoot must presumably be latent in many plant tissues. We shall not deal with the refined laboratory techniques by which shoots can be obtained from tissue cultures but only with cases in which non-axillary development occurs naturally or can be provoked by simple procedures of a more horticultural character.

The appearance of a bud in any unusual position, upon a leaf or a root or an internodal portion of stem, is always of interest. Expressions such as 'regeneration' and 'adventitious bud' have long been current, but their use merely obscures the fundamental diversity of the phenomena. There is no standard pattern of 'regeneration' and no agreement upon the definition of the word itself. One thing which is perfectly clear is that in dealing with developmental processes which can be grouped together only because they are in some degree 'abnormal', one must be prepared to find complex interactions between the controlling factors. Abnormal behaviour is variable behaviour; this is a field of study in which a result once placed on record by a conscientious observer is not discredited simply because some later worker fails to duplicate the effect. This principle has not always been respected; for instance Ossenbeck (1927) was far too ready to dismiss the work of her predecessors on the strength of (sometimes quite casual) negative observations.

The origin of shoots in unusual positions must in some degree obscure the conventional distinctions between one kind of organ and another. Thus Isbell (1931) took leaf-cuttings of tomato, whole tomato leaves from which all axillary bud tissue had been removed. These were planted in sand and took root from the wounded end of the petiole. New shoot buds later appeared in a pseudo-axillary relationship to the upper leaflets. Plants obtained in this way were grown on to flower and fruit; the connection between the upper shoot system and the roots was however by way of the rachis and petiole of the original compound leaves, these organs enlarging and behaving in all respects like the lower part of the stem in a tomato plant grown from seed. Such plasticity of behaviour has very commonly a capricious taxonomic distribution. Potato leaves, although structurally similar to tomato, showed under Isbell's conditions a negligible ability to produce new growth.

In some angiosperms roots and/or shoot buds can be obtained from cut pieces of a leaf which would not have shown any such development if left intact. In some cases (e.g. *Begonia*) these effects form the basis of regular horticultural practice. Regeneration from leaf fragments is generally subject to a decided apical/basal polarity and is governed further by the distribution of vascular tissue. Cutter (1962), working with *Zamioculcas*, obtained regeneration where the largest veins of the leaf-piece, followed towards the base, meet the cut edge. A piece cut as in Fig. 1 (see p. 4) regenerates at the ends of lateral veins, but in a piece cut as in Fig. 2 regeneration is at the midrib only, showing that the minor reticulate venation (not figured) must be effective in conveying a stimulus. Polarity in *Zamioculcas* is strong; a leaf-cutting which is planted the wrong way up still regenerates only at its proximal end. Cutter applied various hormones in paste to the distal ends of cuttings. The effect is in general to increase progressively the amount of regenerative activity, first by inducing growth at such positions as X in Fig. 2, then by causing moderate amounts of regeneration upon the distal margin. A significant reversal of polarity appeared only when cuttings treated with distally applied hormone were also inverted.

The regenerative capacity of a plant fragment must depend upon its size. This is one reason why negative findings are inconclusive. Isbell obtained roots in abundance from isolated tomato leaflets, but unlike whole leaf-cuttings these did not produce shoots. We cannot tell whether this means only that the smaller piece would require more time for shoot initiation or whether the leaflet cutting is under some more fundamental disability. The influence of size and vascularization can be seen in the work of Harris & Hart (1964), who cut squares of lamina from the leaves of *Peperomia sandersii*. Their cuttings were of two sizes, 5×5 mm and 10×10 mm, and were taken in such a way that some were crossed centrally by one of the principal veins of the leaf while others were 'veinless' (i.e. contained only subordinate vascular strands). Squares were subjected to various hormone treatments; we consider principally an experiment involving twenty-four hours of exposure to idolylbutyric acid (IBA) in concentrations of 0, 1, and 10 mg/litre, followed by incubation of the squares. The work differed from that of Cutter in that the application of the hormone was not localized in any way. Buds were slower to develop than roots, and Harris & Hart discussed the two organs separately, a decision which in some respects obscures the significance of their results. For our purposes it will be convenient to take the total regenerative capacity in a batch of cuttings as: (roots per square at day 12 of incubation + buds per square at day 39 of incubation). Recalculating

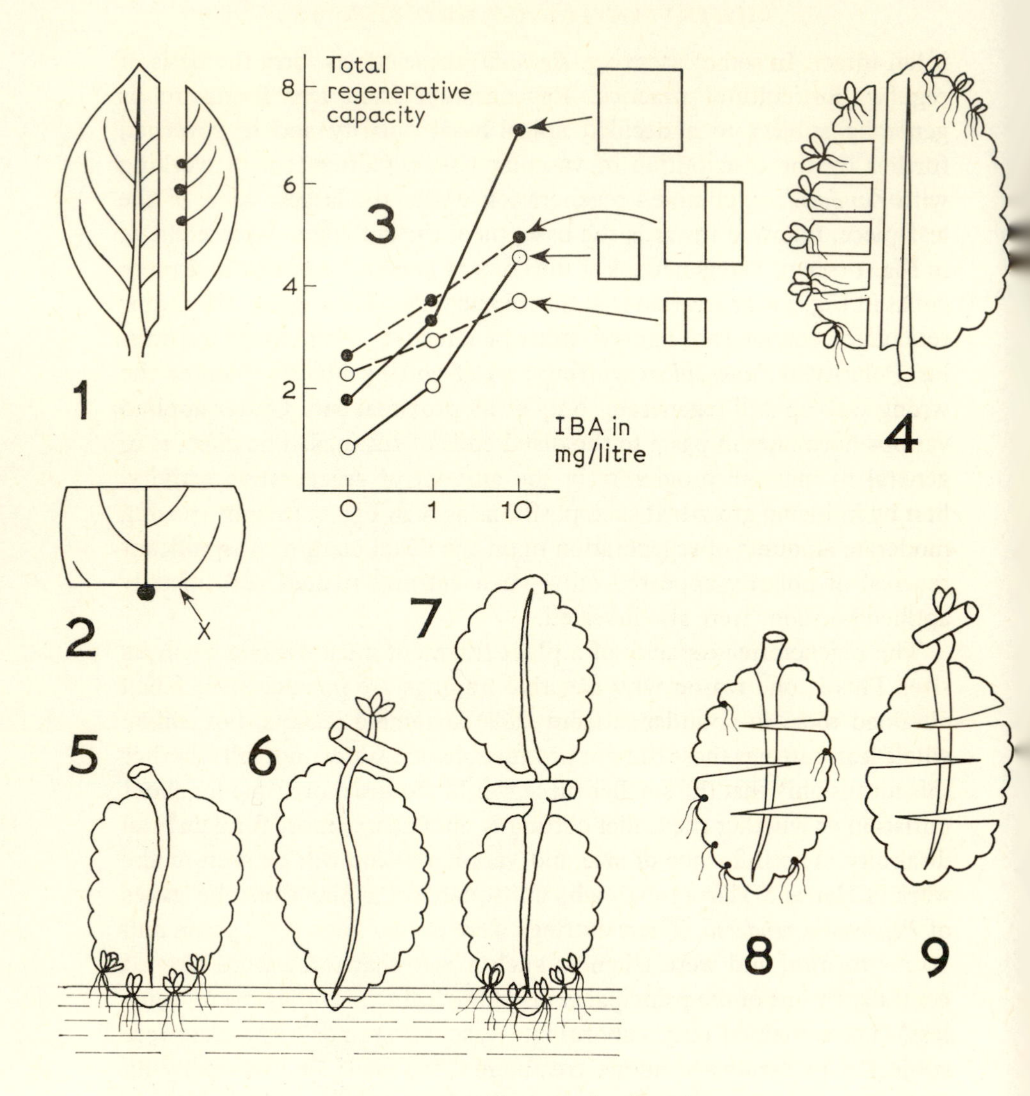

FIGURES 1–9. *Regeneration from leaves.*

1 & 2. Sites of formation of regenerative tubers in two types of leaf-cutting of *Zamioculcas*, X being a point at which regeneration may be induced by chemical stimulation (Cutter, 1967). **3.** Effect of indolylbutyric acid on the regenerative capacity of four types of square leaf-cutting (large/small, veined/veinless) in *Peperomia*. Calculated from observations by Harris & Hart (1964). **4.** Experiment by Loeb (1915) with *Bryophyllum*. Right side shows apical polarity of regenerative activity, while the left side has been divided, destroying the polarity system. **5–7.** Further experiments by Loeb with leaves in a damp chamber, their tips dipping into water. **5.** Isolated leaf regenerating at tip. **6.** Leaf attached to nodal portion of stem, opposite leaf removed; axillary bud of missing leaf conse-

the observations in this way we obtain Fig. 3, from which it appears that: (*a*) at every concentration of IBA the regenerative capacity of a large square is consistently 145% or 150% of that of a small one; (*b*) veinless squares in the absence of IBA have a lower regenerative capacity than veined ones, *but* (*c*) veinless squares are much more sensitive than veined ones to the action of IBA, *so that* (*d*) maximum regeneration is found in veinless squares treated with the highest concentration of IBA. This is a fairly simple set of relationships, yet it leads very obviously to a whole new series of problems. We see, for instance, that with squares of this order of magnitude the regenerative capacity of a leaf-cutting is not, as one might expect, proportional to the square of its linear dimension, but more nearly proportional to the square root thereof. This would suggest that something which is necessary for regeneration is subject to a high rate of wastage in its polarized movement through the tissue, and this in turn generates the idea that rectangles of equal area, but cut with their longer sides respectively parallel and perpendicular to the median axis of the leaf, ought to have different regenerative capacities. It is evident also from other experiments in the same series that new factors come into play when the IBA treatment is more prolonged. Continuous exposure to IBA at 10 mg/litre totally inhibits the production of buds, although its effect upon root production is merely to reinforce the initial stimulus.

There is in general no compelling need to associate the study of shoot development at all closely with the question of root initiation. The ability to produce roots is much more widespread than the power of shoot initiation, the two processes differ greatly in their hormonal responses, and it is doubtful how far it is profitable to think in terms of an organized balance between them. Dore (1955), for instance, found that pieces of *Armoracia* root not less than 2 cm long formed buds and roots in roughly equal numbers, whereas on shorter cuttings the bud: root ratio increases sharply. The observation is interesting, but it is not clear that the introduction of a ratio concept will significantly enrich the

---

quently active, the growth of this bud inhibiting the marginal meristems. **7.** As in **6**, but with both leaves present, axillary buds therefore dormant, marginal meristems free to develop. **8** & **9.** Leaves in damp chamber, analogous to **5** & **6**, but lamina with deep overlapping cuts, illustrating Loeb's claim that inhibition from an active lateral branch can reach the tip of the opposed leaf by a circuitous route. Regeneration in **8** (here in early stage) shows some loss of apical polarity but greatly exceeds that obtainable in **9.**

the study of either organ considered separately. Many such distinctions appear to be differences of timing rather than amount, as exemplified in Table 1 on p. 13.

The largest connected body of knowledge concerning the polarity and spatial distribution of regenerative effects is based upon the species of *Bryophyllum*, usually treated as an independent genus but sometimes included in *Kalanchoë*. The leaves possess distinct patches of meristematic tissue in the notches of the margin. Potentiality for growth is evident here at an early stage in the development of the leaf, and must probably be regarded as a direct continuation of the meristematic condition of the shoot apex. The site of regeneration being fixed by this measure of previous morphological differentiation, the phenomena are on a rather different footing from those observed in leaf fragments of other plants. The species of *Bryophyllum* differ considerably in their reactions. In some the marginal buds of leaves grow into effective units of propagation, with leaves and roots, and are dispersed independently of the parent leaf, which remains attached to the stem. In others the continued growth of marginal shoots in nature is observed mainly upon fallen leaves, the buds upon attached leaves remaining dormant unless stimulated, with varying degrees of difficulty, by specific experimental procedures.

The fundamental problems of *Bryophyllum* regeneration were clearly outlined in the early work of Loeb (1915), who used *B. calycinum*. In this species there is no growth from an attached leaf, but a detached one produces shoots, and often does so with a pronounced polarity. Commonly the shoots arise from notches in about the apical two-thirds of the leaf, the more basal notches remaining inactive. Loeb cut a leaf as in Fig. 4 and claimed that the shoots on the divided side no longer displayed significant polarity, the basal notches having been freed from the correlative inhibition which would otherwise have been imposed by their more advantageously placed neighbours.

In another form of experiment leaf specimens were hung in a damp chamber with their tips dipping into water. Fig. 5 is simply a detached leaf, showing normal apically polarized regeneration. Fig. 6 is one leaf of a pair, still attached to a bit of stem (one node only, the opposite leaf of the pair, but not the axillary bud of that leaf, being removed). In this case there was no regeneration. Loeb's interpretation was that removal of the opposing leaf liberates the axillary bud at the top of the specimen; that bud, in its growth, exerts an inhibitory influence which suppresses the activity of the marginal buds. Fig. 7 shows a corresponding situation but without removal of the opposing leaf. The axillary bud, being retarded by its own axillant leaf, does not grow, and there is no suppression of the marginal buds.

In the picture so presented there are several ideas which call for further consideration. We see first that regeneration is subject to a polarity which can be largely destroyed by cutting the leaf into pieces. This has been questioned by Ossenbeck (1927), who incubated specimens in very wet conditions upon sand, and claimed that any marginal meristem will grow if enough water can be got to it, whether the leaf be cut up or not. Even if polarity is not always apparent, however, there are enough later observations supporting Loeb on this point to establish the reality of the phenomenon. The principle that growth of an axillary bud may be released by removal of its axillant leaf is well authenticated and calls for no special comment. The most controversial part of the Loeb scheme is the inhibition of marginal buds by a small axillary shoot at a considerable distance from them. Loeb was firmly convinced of the reality of such an inhibition, and sought to show, by setting up the experiments of Figs. 8 & 9 (analogous to Figs. 5 & 6, but with deeply overlapping transverse cuts in the lamina) that the suppression was capable of taking a very indirect course.

Loeb in his later writings alienated the sympathies of many readers by introducing physiological hypotheses of quite unacceptable crudity. His ideas upon the possible effects of gravity were particularly vulnerable to ridicule, being based upon 'accumulations of sap' in a manner reminiscent of the seventeenth century. So it was perhaps not in a completely impartial frame of mind that Ossenbeck suspended detached leaves in a damp chamber with surface vertical and midrib horizontal, and observed the relative activities of the buds upon the upper and lower margins. In the event, the percentages of preformed buds which were stimulated into further growth were for *B. calycinum* 57% upper: 39% lower, for *B. proliferum* 32% upper: 52% lower, for *B. crenatum* 66% indifferently upon either edge. The number of observations is only approximately indicated in the original text but seems not to have been absurdly small; one might have thought that this was an inconclusive but reasonably promising investigation, likely if pursued to reveal genuine differences in gravitational response between the three species. Ossenbeck in her summary merely states flatly: 'no gravitational effect could be demonstrated'. Against such a background every positive result would appear to have some value.

A significant advance was made by Dostál (1931), who used *B. crenatum* and obtained quantitative results by determining at the end of each experiment the dry weights of the leaf-cutting and of the new organs (buds and roots) produced from it. His measure of productivity was: (dry weight of new organs in mg per g dry weight of original leaf). The values obtained naturally depended on the duration of the experiment, which seems generally to have been from four to six weeks but was not standardized. Commonly the two leaves of a pair were used for different treatments. Between different series of results there was no real comparability in Dostál's original presentation; to some extent we can get over this difficulty by further calculation. It must be noted also that there was an important distinction between subdivision of a leaf at the beginning of an experiment (to see how the buds would behave) and subdivision at the end (to ascertain the relative productivities of different parts).

It appeared to be a rather general rule that the apical part of a detached leaf was more productive than the lower part, and that total productivity could be increased, up to a point, by cutting the leaf into pieces. Figs. 10 & 11 (see p. 9) show two leaves of a pair, one incubated whole, the other first divided into quarters. Actual productivities (in mg/g) are given for the upper and lower halves of each leaf. Division has markedly increased the general level of productivity, but reduced the bias in favour of apical buds. All such results are somewhat precarious, in that any defect in the conditions may prevent the higher productivities from being realized. It is not altogether reasonable to expect that such experiments should be repeatable at will.

Dostál went on to try the effect of cutting leaves in more complicated ways. Some of his manipulations are shown in Figs. 12–17. The productivities placed upon the various pieces are relative values, calculated upon the convention that a productivity of 100 shall be attributed to the portion which is attached to the petiole. It must be remembered also that productivity is measured for the whole of each piece, so that considerable caution is needed in making comparisons. There is no reason to think, for instance, that the apical part of the leaf in Fig. 12 is any less productive than the corresponding part in Fig. 16. There appears, however, to be a very general confirmation of the existence of polarity, and a clear indication, in Figs. 14 & 15, that the contribution which a piece of leaf tissue can make to regenerative productivity depends upon its ease of access to the margin.

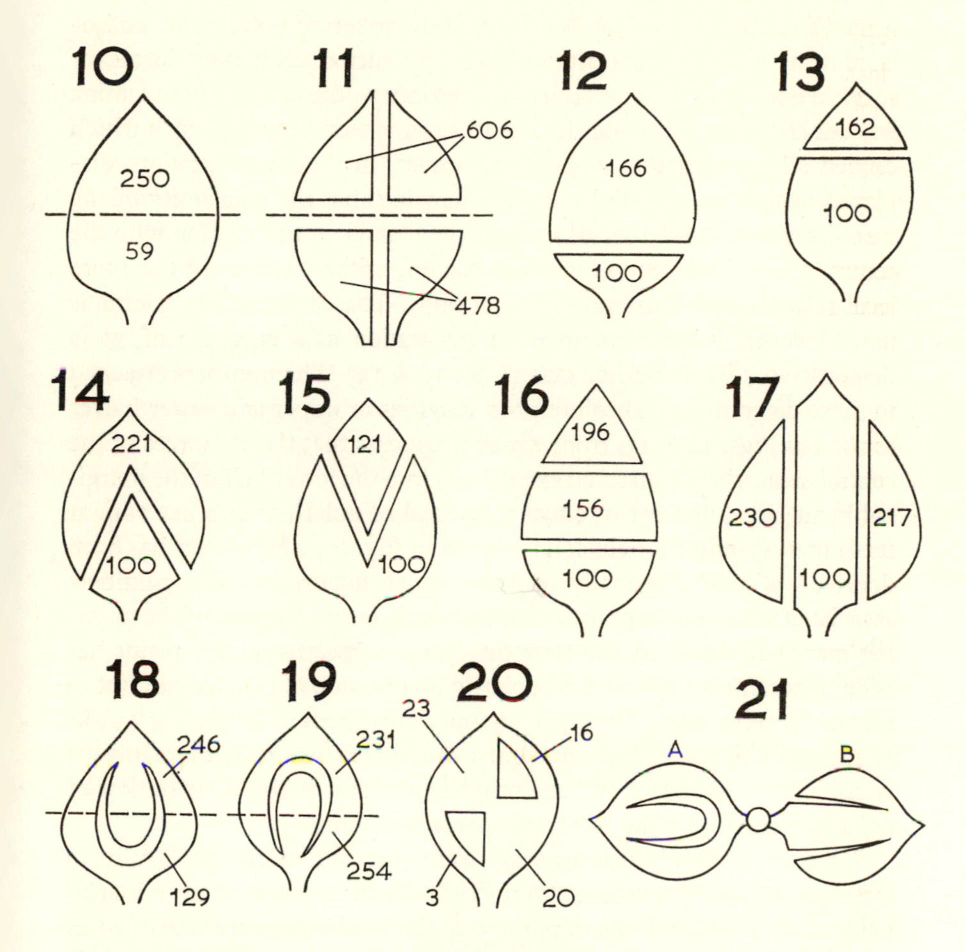

FIGURES 10–21. *Surgical procedures in the study of regeneration from leaves of* Bryophyllum.

Cuts made at the beginning of an experiment are shown in full; a broken line denotes separation of parts only in the final process of measurement. **10 & 11.** Observed regenerative productivities of apical and basal halves of leaves incubated whole and in quarters (Dostál, 1931). **12–17.** Further experiments by Dostál, productivities of the pieces on a relative basis, the portion attached to the petiole taken as 100. Examples in the bottom row are from Uhrová (1935). **18 & 19.** Observed regenerative productivities of apical and basal halves of leaves with horse-shoe cuts. **20.** The relative sizes of regenerated shoots in the four quarters of leaf deprived of quadrants of central lamina. **21.** A pair of leaves cut so as to redirect the flow of extrafoliar inhibition (see p. 11).

Uhrová (1935), after taking some measurements which merely confirmed the observations of Dostál upon the inherent polarity of the isolated leaf, made significant advances by introducing new forms of surgical operation. In one series of experiments the centre of the lamina was cut right out, reducing the leaf to a frame 2 or 3 mm in width which carried all the marginal buds. Such leaves, in their regeneration, displayed normal apical/basal polarity, showing that the continuity of the vascular system can be seriously disrupted, and communication with the central area of the lamina entirely cut off, without affecting the more local relationships between marginal buds. The centre of the lamina is not, however, inoperative in the regeneration of a normal leaf, as is demonstrated by horseshoe cuts (Figs. 18 & 19). The numbers attached to these diagrams are absolute productivities of upper and lower halves as used in Figs. 10 & 11, from which it appears that the resources of the central area, when redirected exclusively into the lower half of the marginal band, are sufficient to effect a reversal of polarity. Another variant tending to a similar conclusion is shown in Fig. 20, where a leaf has been deprived of two diagonally opposite inner quadrants. The numbers associated with this diagram are average-size measurements of the growing marginal shoots in the four quadrants respectively; the result has been to accentuate the natural polarity on one side of the midrib, but to reverse it on the other. By repeating this experiment with varying widths of marginal tissue in the operated quadrants it would presumably be possible to build up a rather complete picture of the distribution of polarity control over the entire surface of the lamina.

Some of Uhrová's experiments were directed more specifically to interfere with any communication which might originate from a marginal bud. One method was to remove all the buds along one side of a leaf except for one selected bud. This chosen bud, unless it is in a decidedly basal position, will then grow more actively than its counterpart on the unoperated margin of the same leaf. This result establishes a strong presumption that each marginal bud exerts some inhibitory influence upon the others. Another mode of operation was to take the two leaves of a pair and to retain upon both a small number of morphologically equivalent buds in the middle region, while completely disbudding the apical part of one leaf and the basal part of the other. It is then found that the buds at the middle grow larger in the apically disbudded leaf than in the basally disbudded one. This tends to show that transmission of the mutual inhibition between marginal buds is more effective in the basipetal direction. Uhrová seems curiously to have missed the point that these disbudding procedures, like the selection of a bud, might instructively be applied to a single margin.

So far as polarity of the leaf is concerned, the work so far reviewed may be considered indicative of an interaction between several factors. Marginal buds emit correlative inhibitions, the strength of such an inhibition at source being presumably a function of the bud's developmental condition, very possibly also of its position on the leaf. The inhibitions are transmitted preferentially, though perhaps not exclusively, in the basipetal direction. By these and/or other means the marginal region can maintain an apparently normal polarity without assistance from the rest of the leaf. The local arrangements in the marginal tissue are however seriously disturbed when the communications with the central area, instead of being totally cut off, are diverted into abnormal channels. It therefore seems unlikely that polarity in the intact leaf can be entirely independent of the central lamina.

The study of the inhibition imposed upon the marginal buds of a leaf by the attached organs (petiole, node, axillary buds, etc.) must be rendered more difficult by the complexity of the internal situation in the leaf as just outlined. Any inhibitive influence of extra-foliar origin enters upon an already crowded stage. As to the existence of such an inhibition, in those species where marginal buds do not develop in the intact leaf, there can clearly be no dispute, but it is not surprising that authors should disagree as to its strength and the effectiveness with which it is transmitted. Loeb found a strong and pervasive inhibition from very small portions of attached stem, while Ossenbeck and others were less impressed with the importance of this phenomenon. Uhrová made an original contribution by setting up the experiment shown in Fig. 21, where cuts have been made in a pair of leaves, still connected to each other by their petioles and a short nodal portion of stem, but separated from the rest of the plant. The cuts are so placed that any inhibition coming from the stem will be directed, in leaf A, into the basal part of the lamina, and will be able to reach the apex only by a circuitous route, while in leaf B any inhibition from the stem will be channelled to the leaf-tip with the least possible opportunity for diversion. It is further to be noted that the surgical procedure near the tip of the leaf, and for some distance downwards, is the same in both leaves. This ingenious manipulation produced a positive result: the most active buds upon leaf A, and the least active upon leaf B, were precisely those in line with the apical ends of the incisions, which could hardly be due to any cause internal to the leaf.

Some of the experiments carried out by such investigators as Loeb, Dostál, and Uhrová may be regarded as attempts to discover just how much of the rest of the plant must be removed from a leaf to release the growth of the marginal buds. An alternative approach is to endeavour to render the inhibition inoperative by other means, short of actual amputation. Ossenbeck (1927) and Mehrlich (1931) both give interesting lists of treatments recommended by previous workers which they had tested with varying measures of success. In general these procedures become more informative as they become more remote from physical separation of the leaf. That it should be possible to induce the development of buds by constricting the petiole or stem with clamps or tight bindings of thread is hardly surprising; that it is possible to do so by cooling the petiole to the (apparently non-damaging) temperature of 4°C. is a more substantial addition to knowledge, while the fact that it is (sometimes) possible to induce marginal growth by keeping the plant in the dark may well be a major clue to the underlying physiology.

The value of much of the older work is diminished by the uncertainty of some of the methods employed. It is not always possible to ascertain the conditions under which plants were exposed, for instance, to atmospheres enriched with hydrogen or carbon dioxide, nor is it clear just what is involved in setting a shoot apex in a plaster cast. One thing which plainly emerged, however, was the existence of significant taxonomic differences. From the work of Ossenbeck one can prepare Table 1, in which three species are shown to be clearly distinguishable by their regenerative behaviour. Mehrlich went further; working with stocks of *Bryophyllum calycinum* from five different sources he found that some of his races would proliferate from attached leaves when placed in the dark, whereas others would not.

Later experimental studies upon those species of *Bryophyllum* (particularly *B. tubiflorum*, *B. daigremontianum*, and *B. hybridum*) in which buds normally become active upon attached leaves have concentrated upon the problems of photoperiodic response. It was shown by Götz (1953) that the development of the marginal meristem up to a certain stage was independent of day-length. Beyond that point there was no further growth in a regime of short (nine-hour) days but an attached leaf (not too old) would go on to produce and shed normal bulbils if given twelve and a half hours of light for nine days. Regardless of day-length, removal of the leaf from the plant at once induces growth of the buds. Kröner (1955) has found that a nine-hour day, though severely unfavourable to bud development, is not absolutely prohibitive. In time, and with high light intensity, plants under a nine-hour regime ultimately produce bulbils, though those with only seven hours do not. Kröner has revealed another complication by planting successive batches of bulbils from January to April, keeping them all in nine-hour days until June, and then transferring them all to a sixteen-hour regime. In this experiment the February and March plantings showed the quickest response.

TABLE I

Differences in regenerative behaviour between three species of *Bryophyllum* (observations by Ossenbeck, 1927)

	B. crenatum	*B. calycinum*	*B. proliferum*
Sequence of organs from foliar buds	Shoots first, then roots	Roots first, then shoots	Shoots, then roots
Axillary buds	Two unequal buds in each axil, smaller only occasionally growing when larger not removed	One bud in each axil	Two unequal buds, smaller never seen to grow unless larger removed
Regeneration from roots	No buds	No buds	Copious formation of adventitious buds
Response to low temperature	Leaf buds become active if root system chilled to 0°C. for a few days		No response

The January batch was very slow to respond; after more than two months in long days the January plants were still markedly behind the April ones (half their age), so lasting was the effect of prolonged exposure to the nine-hour day. Götz and Kröner, though not in agreement upon details of procedure, both appear to have demonstrated the existence, in sap extracted from short-day plants, of a specific chemical inhibitor, whereas the sap of long-day plants is not known to contain any substance which can stimulate the formation of bulbils. Götz found that different regions of a shoot, when subjected to long and short days respectively, did not affect one another; this result is consistent with the general failure to obtain significant changes of behaviour by grafting experiments. In the course of these investigations it has incidentally become clear that an additional inhibitory factor exists in the inflorescence region, depressing the activity of the marginal meristems from the first appearance of flower-buds until the seeds have ripened.

In all physiological studies on regenerative phenomena one must be prepared to find evidence of interaction between the agents applied; it is quite characteristic, for instance, that Ossenbeck should find buds on the roots of *Bryophyllum proliferum* developing only when illuminated, unless supplied with nutritive solutions. Given an enhanced level of nutrition, light ceased to be necessary. It is necessary also to distinguish between different stages of development of the bud, because the requirements for the initiation of an apical meristem may be entirely different from those for continuation of its activity. Thus Williams, Dore & Patterson (1957) found in *Armoracia* root that the impetus to develop shoot primordia was remarkably strong and that initiation took place in conditions (lack of oxygen, or abnormally high temperature) which completely inhibited further progress. Another point which has to be taken into account is the possibly circuitous nature of some of the effects observed. Ossenbeck was able to bring about the growth of marginal buds in *Bryophyllum crenatum* and *B. calycinum* by etherization of the plant. This did not, however, appear to be a direct action upon the buds themselves or upon any part of the leaf, because etherization of a single leaf proved ineffective. The presumption must rather be that the narcotic depresses the activity of the apices already growing, thereby interrupting the correlative inhibition which these apices exert upon the marginal meristems.

It has also become clear that the conditions required for regeneration, and indeed the very possibility of obtaining regenerative development, are dependent on the age and morphological status of the material. Dore (1955), in dealing with *Armoracia* root, used the expression 'ripeness to regenerate'. For the production of shoots from his roots it was necessary that (*a*) the primary cortex should have died; (*b*) the original pericycle should have developed as a phellogen. These conditions were satisfied at the age of one to two months. Subject to further conditions (including separation from the plant) shoot buds would then appear. At the time of Dore's investigation there was already an existing literature dealing with the regeneration of much older roots of the same species. It appears that this regeneration from really old roots must be regarded as an altogether different process. The anatomical features are quite distinct, the polarity of the older root is much more pronounced than that of the younger one, and the physiological requirements are not the same.

The condition of the *Bryophyllum* leaf, with active shoot production from marginal meristems which can be regarded as residues of the parent apical meristem, is viewed by Stoudt (1938) as the end-product of an evolutionary sequence, the beginning of which is to be sought in other Crassulaceae with more generalized forms of foliar proliferation. To see what Stoudt regards as more primitive patterns we may consider the work of McVeigh (1938) on *Crassula multicava* and that of Yarbrough (1936b) on various species of *Sedum*. In none of these plants does the leaf contain any residual meristem or make any move towards the production of buds while it is undamaged and still attached to the plant. A detached leaf, however, produces shoots and roots at its basal end. In *Crassula* the ability to regenerate is generally distributed through the leaf; an apical half freely produces buds and roots, showing the same kind of polarity as Cutter observed in *Zamioculcas*. In *Sedum* a leaf generally produces only a single shoot, and although regeneration from apical pieces has been reported it is clearly much easier to obtain a bud from the region of petiolar attachment than from any other part. The anatomical origin of buds is quite different in the two genera. In *Sedum* the new organs arise from an active callus, a substantial cushion of parenchyma which forms upon the wound surface. In *Crassula* the new growth, though near to a wound, is structurally independent of the wounded area, and is initiated by periclinal division in fully differentiated epidermal cells. Stoudt selected *Sedum* as his primitive type; in that regeneration of *Bryophyllum* is not closely associated with wound surfaces one might think *Crassula* a more logical choice, and this consideration is somewhat reinforced by the observation that the *Sedum* leaf often has a particularly thin and brittle petiole, and serves regularly as a unit of propagation, whereas the establishment of new *Crassula* plants from fallen leaves seems to be of a more accidental character. Stoudt's own studies were concerned with *Byrnesia* and *Kalanchoe rotundifolia*, where each leaf possesses one residual meristem, not however marginal in position but situated adaxially near the base. Such a meristem, though plainly additional to the customary developments of the axil, is comparable in origin to an axillary bud. Stoudt naturally regards this arrangement as intermediate, and sees the origin of the *Bryophyllum* type as a progressive distribution of the residual meristem into marginal positions and an increasingly precocious display of its activities. Upon this basis, *Bryophyllum* species which shed bulbils from the attached leaf are more highly evolved than those in which only fallen leaves develop new shoots.

In any case of regenerative development it is to be expected that the early stages of growth will be of a morphologically indeterminate nature; several mitotic cycles must generally take place before the distinguishing features of a root or shoot apex can appear. In some instances the site of mitotic activity will indicate the kind of organ which is to emerge. In leaves of Crassulaceae the rule that shoots arise at the surface while roots are endogenous seems very generally to be maintained. This is by no means universally the case, however, nor does the plant invariably proceed by the shortest method to the production of normal root and stem structures. Cutter's *Zamioculcas* leaves regenerated by the production of an endogenous tuber, which reached significant dimensions before other organs began to emerge from it. Endogenous regenerative activity seems rather commonly to involve the production of an appreciable volume of 'uncommitted' tissue. Regeneration necessarily takes an endogenous course in any root or stem old enough to possess periderm, or in any instance of development from a wound surface, and may do so even in organs where the original epidermis survives. Accordingly there must be many cases of regeneration where it will be necessary to consider the possibility that the pattern of differentiation might be freely convertible, as between shoot and root, by appropriate stimulation at a relatively late stage.

When root and shoot apices are arising in close proximity it is natural that there should sometimes appear to be a close association between one root and one shoot. McVeigh, for instance, in dealing with the origin of new plants from pieces of *Crassula* leaf, was able to photograph a 'longitudinal section of a young embryo' which was indeed strikingly similar in general appearance to some of the embryonic stages of ferns and other plants. The *Crassula* 'embryo' had a root at one end and a stem at the other, and a lateral region, attached to the parent leaf, which might reasonably bear comparison with the nutritive foot of the fern embryo. It is not clear, however, that any advantage is gained by imposing a rigid 'embryo' – concept upon the treatment of regenerative developments. There is no question in *Crassula* of any new structure being referable in origin to a single cell of the original leaf. Nor has it been demonstrated that any precise demarcation exists between the new organs and the old; some cells perhaps cannot be assigned with certainty to either. It appears also that the association of one shoot with one root, so necessary to any general rule of embryology, is not constantly maintained in *Crassula*. Often a substantial root system is first established from the leaf-cutting, unaccompanied by any shoot apex at all, and then stem buds appear by a secondary process of initiation upon the larger roots. On the whole then, any conformity of *Crassula* regeneration to the conventional patterns of embryology, though not particularly rare, appears to be of a merely fortuitous nature. Another difficulty lies in the character of the vascular connections; the new organs, whatever independence they may show in their mode of origin, very soon become linked with the vascular system of the parent leaf, and this is very different from the behaviour of any true embryo.

In McVeigh's work the idea of a foliar embryo makes only an incidental appearance, and is treated as a descriptive device, devoid of theoretical significance. It has been given much greater prominence by Naylor (1932) and Yarbrough (1932), whose investigations seem to have been quite independent although so nearly simultaneous. Both workers clearly recognized that the leaf of *Bryophyllum calycinum* possesses in the notches between its crenations residual meristematic patches consisting of many cells, that the ability to grow new organs is found in these meristems and nowhere else in the leaf, and that the stem apex in each notch, which is directed towards the adaxial face of the leaf, is continuous with the general leaf surface. This stem apex, in the mature leaf, already shows at least one foliar primordium (Naylor and Yarbrough both say two, but their drawings appear indicative rather of a single leaf and a residual dome of stem tissue). It is agreed that the roots arise internally in the parenchyma and are directed towards the abaxial face of the parent leaf, but Yarbrough finds that roots become recognizable only after the leaf has been detached, and is non-committal about their number, whereas Naylor finds conspicuous root apices even in attached leaves and attributes a regular complement of two to each 'foliar embryo'. These studies at least establish beyond question the existence in the *Bryophyllum* leaf of a degree of organization which is lacking in *Crassula*. That regenerative power should be confined to the residual meristems, not appearing in other tissues even when those meristems are excised, is an interesting commentary upon Stoudt's evolutionary hypothesis. It is curious also that the tip of each crenation should contain what Yarbrough called an 'apex patch' consisting of tissue very similar in appearance to the residual meristem but without any of its regenerative potential.

The concept of a foliar embryo cannot really be maintained, however, without doing some violence to the facts. Naylor goes so far as to show in his diagrams a smoothly rounded embryo with shoot apex, two root primordia, and foot, although his anatomical drawings prove that this is not a dissectible entity and that his 'foot' is an arbitrarily chosen part of a continuous parenchyma in which no natural boundary can be found. The continuity of the vascular system of the supposed 'embryo' with that of its parent is fatal to claims which exaggerate the independence of the new growth. Yarbrough in his later work tacitly abandoned the rigid application of the foliar embryo concept. In *Tolmiea* (Yarbrough, 1936) he found each leaf to possess a single bud, formed from residual meristem, at the junction of petiole and blade. This bud, unlike those of *Bryophyllum*, develops continuously, so that an adult foliage leaf carries a well-established leaf rosette. Roots are formed much later, and from a quite different part of the parent petiole. In the fern *Camptosorus* (Yarbrough, 1936a) root primordia in considerable numbers are established in the tip of a leaf even before this is unrolled. At a rather later stage, but before the parent leaf tissues have fully matured, a stem apex arises from one of the marginal cells of the growing lamina. From Yarbrough's discussion it is evident that regenerative phenomena in ferns are very diverse. There seems to be no immediate prospect of recognizing any general law regarding the evolution of regenerative mechanisms, but in some instances one can indicate points of apparent ecological specialization. Götz, for example, suggests that the bulbils of *Bryophyllum daigremontianum* and similar species may owe their dispersal to a flicking action, based on the stiffness and springiness of the leaves and of their lateral appendages. Nothing seems to be known of the actual abscission of the bulbils, but a refined dispersal mechanism would be consistent with Stoudt's views on the evolution of the Crassulaceae. An even more distinctive specialization towards the production of standardized compact propagules can be observed in *Malaxis paludosa* (Taylor, 1967). In this orchid the tip of a leaf produces a cluster of small bodies, numbering about forty in a published drawing. Each of these consists of an almost uniform multicellular knob enclosed in a sheath which has an apical pore. The closest comparison seems to lie with the gemmae of bryophytes; the structures in *Malaxis* contain no vascular tissue and are not served by any special extension of the vascular system in the parent leaf. Taylor's account does not fully explain the process of dispersal, but it appears that the gemma travels complete with its sheath, that it does not at the time of liberation contain any fungal mycelium, but that it is readily infected from the soil by the characteristic fungal associate of

Malaxis. In such a case one is naturally disposed to see in the mycotrophic habit of the plant an opportunity, if not an explanation, for the small size of the propagule, for the rather large numbers in which it is produced, and for its lack of anatomical differentiation.

In many plants the hypocotyl is capable of producing shoot buds upon its sides. It appears that in most cases these buds are initiated by cell divisions in the epidermis of the hypocotyl. Link & Eggers (1946) found in cultivated flax that the lower part of the hypocotyl regularly produced a number of epidermal bud-primordia, though in intact plants it is uncommon for any of these to develop very far. Decapitation of a young hypocotyl, involving the complete removal of plumule and cotyledons, stimulates the upper half of the hypocotyl to produce bud-primordia, although it would not normally have done so. Whether decapitation also increases the number of buds in the lower half of the hypocotyl is not known, but the number of potential sites of shoot growth is in any case far in excess of any ecological requirement, as many as fifty having been observed in a single decapitated hypocotyl, and up to thirty-five in the corresponding part of an intact seedling. The ability to initiate the process of shoot formation in this species is apparently a prerogative of rather young epidermal cells. After a seedling has reached a certain age decapitation will no longer induce the upper part of the hypocotyl to initiate primordia. For some time after the ability to generate new primordia has been lost, however, decapitation continues (though with a declining rate of survival) to be effective in provoking the further development of primordia already in existence. Probably because of correlative inhibition, it is not usual for more than eight or ten buds to reach the stage of obvious leaf-formation, and one shoot generally attains total domination at an early stage. The hypocotyl of *Vaccinium macrocarpon*, which has been examined by Bain (1940), has not been seen to produce, in the intact seedling, anything which could be regarded as a bud-primordium. If decapitated, however, it produces a small number (about five) of buds in its upper part, and from these a viable plant can be established.

In both *Vaccinium* and *Linum* the first stage in the production of a bud-primordium is a transverse division in an epidermal cell of the hypocotyl. In *Vaccinium* and usually also in *Linum* there are other transverse divisions, giving a vertical row of cells within the elongated outline of the original cell. Further divisions take place in both radial and tangential longitudinal planes, and as there is no very rapid increase in the total size of the product tissue the individual cells become very small. Some of the hypodermal tissue also becomes involved, but for a long time there is no visible effect on the general mass of the cortex. Only if the primordium at the surface is stimulated into the development of leaves do the inner tissues of the hypocotyl differentiate so as to provide the new bud with a vascular connection. The bud-primordium in its characteristic resting state is therefore an isolated and essentially superficial piece of meristem; no stem apex can be distinguished in it at this stage. The primordium does not appear to be derived from a single cell, nor would it be accurate to describe it as purely epidermal, but it does seem to be true in both species that only one of the original epidermal cells becomes involved. The other cells of the epidermis are merely pushed aside in a way which can misleadingly suggest an endogenous origin for the bud.

In many dicotyledons belonging to various families, and more exceptionally among monocotyledons and pteridophytes, the ability to produce shoots from the root system and hypocotyl plays an important part in determining the life-form and ecological capacities of the species. Rauh (1937) has demonstrated the existence in the genus *Euphorbia* of a series of life-forms which may be arranged as follows:

E. helioscopia is not known to develop buds below the cotyledonary node in any circumstances.

E. bubalina produces buds from the hypocotyl, but only as a response to injury.

E. segetalis regularly produces hypocotyl buds, which grow out and bear flowers.

E. gerardiana produces buds both from its hypocotyl and from its roots; any of these may bear flowers.

E. cyparissias produces no flowers from the shoot system above the cotyledons, nor from the branches arising on the flanks of the upper hypocotyl. Reproduction is confined to adventitious shoots arising from root and lower hypocotyl.

Sequences which are less complete, but not different in principle, can be found in other genera such as *Linaria*. Perennial species which can produce buds from lateral roots are well equipped to survive as weeds of arable land. *Cirsium arvense* and *Convolvulus arvensis* come into this category and agree in the following features: the shoot system above the cotyledons dies without flowering; there is a richly branched lateral root system with shoot buds upon it; these buds give rise to aerial shoots which commonly branch from their lower axils. *Sonchus arvensis*, though significantly different in its early stages, ultimately attains a similar condition.

Propagation from buds on spreading lateral roots is not very different in its ecological consequences from the development of stolons or rhizomes. *Cirsium arvense* gains an obvious advantage over its near allies, all of which are biennials with no significant ability to spread by vegetative means, but Rauh cites pairs of related species in which the adoption of root propagation is not associated with any conspicuous change in the general habit of life. *Ajuga reptans* spreads by stolons whereas *A. genevensis* has no stolons but instead produces shoots from shallow-running lateral roots which are very unlike the deep-going absorptive roots; the ecological consequences of this distinction appear to be minimal. A similar pair consists of *Sium latifolium* which grows up from roots and *S. angustifolium* which does not. There seems to be considerable force in the argument that root-propagation and rhizome/stolon-propagation are mutually exclusive alternatives. Some species with a markedly gregarious habit of growth owe this characteristic to development from lateral roots (e.g. *Epilobium angustifolium*). Portions of the system are then very commonly separate at maturity owing to death of the intervening lengths of root; *Rumex acetosella* is an example of the contrary, with shoots over a considerable area organically connected and liable to die simultaneously when part of the area is affected by fire or herbicides.

In some cases of extreme ecological specialization the development of the primary shoot is negligible; the root pole of the embryo becomes dominant from an early stage, and all parts of the adult sporophyte are derived from the root and/or hypocotyl. This is the common course in such parasites as *Orobanche* and the Balanophoraceae, where the cotyledons and the plumular area remain within the seed-coat and are purely haustorial. The hypocotyl forms a tuber, from which other haustorial structures possibly equivalent to roots then enter the host. With various differences of detail the flowering shoots arise from this system. *Monotropa*, though parasitic on its mycorrhizal fungus rather than another angiosperm, has a similar history. In a quite different ecological context the Podostemaceae are small hydrophytes closely attached to rocky surfaces; their seedlings show an early abortion of the plumule, and the growth of a thallus which appears to be a flattened photosynthetic root system of a highly specialized type. The shoots of the adult are almost wholly devoted to reproduction, and arise upon this thalloid root, which may be variously lobed but does not directly give rise to leaves.

The systematic survey undertaken by Rauh has shown that the production of shoot buds by the part of the plant below the cotyledonary node is not, in dicotyledons, a particularly rare phenomenon, and it indicates clearly that many families must independently have achieved varying degrees of evolutionary progress in this direction. At one end of the scale propagation from root or hypocotyl is merely latent in the uninjured plant or may be limited to the production of reserve buds which remain dormant until the main shoot system is destroyed; some trees possess this kind of organ reserve. At the other extreme the primary shoot system is no more than a theoretical abstraction. Where it is possible to make close comparisons between related forms the evidence seems to suggest that there is no fundamental distinction between budding from the hypocotyl and budding from roots. Rauh considered it to be a rule that budding starts in the upper hypocotyl and spreads downwards into the root system, and further saw a link between the anatomical origin of a bud and its timing. An early bud would be in the epigeal part of a hypocotyl, and might be exogenous, while a late one would arise on a root and would necessarily be endogenous. The associated structural details can be generalized only imperfectly; buds upon a root system, for instance, are usually found at the points where lateral roots have emerged, but *Sonchus arvensis* does not conform to this rule.

The buds in the axils of the cotyledons have been found, especially when multiple, to occupy a special place in the life of some species. Rauh studied *Linaria cymbalaria*, in which the foliage and flowers of the first season are produced from the plumule and from the principal bud of each cotyledonary axil. These shoots are lost at the end of the year and the plant has been described in floras as an annual. In reality however it is often kept active for a second season, sometimes even longer, by the growth of new shoots from the numerous subordinate buds in the cotyledonary axils. In a few Papilionatae the seedling (Dormer, 1945a) has no vestige of a plumule, the shoot system being derived usually from four buds (the two largest from the zigzag sequence in the axil of each cotyledon). The subject is one which has been sufficiently neglected to make it likely that other such cases remain to be discovered.

In many trees, almost all in tropical or subtropical climates, flower production is concentrated in short lateral branch systems which burst their way out through the mature bark of the older stems. This cauliflorous habit is conspicuous in rain forests and occurs in some economically important plants such as cocoa. It has not however been sufficiently investigated from the anatomical point of view. Formerly it was supposed that the shoots in question originated from axillary buds, surviving from the sapling stage, which had somehow become embedded in the bark of the trunk and larger branches. Such detailed studies as are available (McLean Thompson, 1951, 1952) show this to have been a mistake in some cases at least. Flowering shoots can arise under the bark in a truly endogenous manner, with no ascertainable relationship to the distribution of nodal features on the younger stem. Similar considerations probably apply in those instances, some of them familiar in temperate latitudes, where vegetative twigs sprout from old bark-covered surfaces.

Chapter Two

The Measurement of Growth

Provided that an organism is protected from sudden disturbances, and provided that observations are taken at intervals which are not excessively short, growth appears to be a smooth and gradual process. A growth curve, which is a graph showing how size increases with time, is therefore of apparently simple form, and the widespread general agreement in shape between growth curves derived from biologically diverse sources has induced many writers to fit equations to the observational data. Such enterprises are customarily based on some simple assumption concerning the quantitative regulation of growth. The assumptions that can reasonably be made are few in number and wide in their field of application; we find therefore that an extensive literature (relating to animals as well as plants, and to problems of mortality and survival as well as the growth of the individual) is dominated by a restricted set of elementary mathematical ideas.

The historical development of algebraic formulations for processes of growth has not followed a strictly logical course, but in the comparative study of growth equations at the present time it can be seen that those which can lay some claim to logical justification have all been founded upon a single principle: that the growth rate of an organism should be related to its existing bulk. Let x be the dimension under observation, t the elapsed time, and k a constant. Then in the notation of the differential calculus we may write:

$$\frac{dx}{dt} = kx \qquad (1)$$

(absolute growth rate proportional to size attained) or the equivalent form:

$$\frac{1}{x}\frac{dx}{dt} = k \qquad (2)$$

(relative growth rate constant). This is the compound interest law, the application of which to the growth of the higher plants has been advocated by Blackman (1919). It is the situation which would exist if every part of the body increased by a fixed percentage per unit time, independently of all the other parts. This consideration is sufficient to expose the fundamental artificiality of the compound interest scheme; it represents the growth of a body which remains meristematic throughout. The expression 'compound interest' is here also used in the special sense of continous compounding. In financial transactions the addition of interest to the capital is made only at stated intervals; interest which falls due does not itself begin to earn interest immediately, but only after it has been added to the capital at the next audit. Obviously the final yield will be the greater as the audit dates are more frequent. In the growth of a plant every bit of interest must be 'capitalized' the moment it is earned. It is this need for continuous compounding which is responsible for the general occurrence in growth equations of natural logarithms and of e, their base number. Integration from either of the equations (1) and (2) above will give:

$$\log x = \mathrm{k}t + C \qquad (3)$$

where C is a new constant, the constant of integration. The value of C is indeterminate (see Fig. 22 and caption) because no zero point has yet been fixed for the scale of time. The neatest way of settling the matter is to introduce x_0, the value of x at zero time; any possible value of x can be chosen for this distinction, at our own convenience. We have then:

$$x = x_0 \mathrm{e}^{kt} \qquad (4)$$

This equation undoubtedly expresses certain biological realities. It indicates, for example, the importance of a good start, because x_0 affects the value of x at every stage. Blackman was able to illustrate this very strikingly from *Helianthus*, where varieties which differ enormously in adult stature have similar percentage growth rates (i.e. similar values of k) but very different seed weights (i.e. different values of x_0). Among these plants, the individual starting from a small seed suffers a proportionate handicap throughout life. Compound interest cannot however give a complete account of the growth of any organism. It implies growth without any check or retardation; no part ever ceases to contribute to further enlargement, and final size is always infinite. Such things are contrary to experience, and compound interest serves mainly as a useful approximation in dealing with young organs and early stages of development.

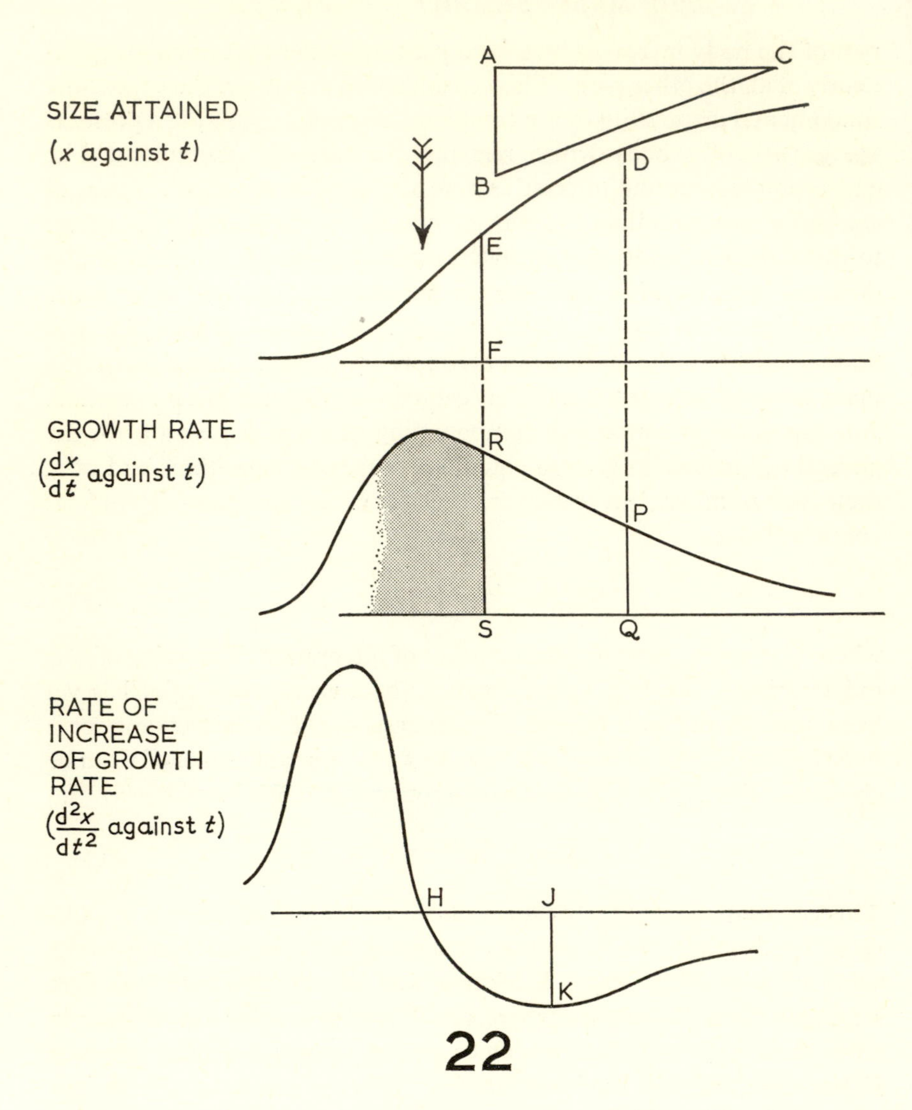

FIGURE 22. *A sigmoid curve, with its first and second differentials, plotted on a common time-base.*

To derive a graph lower on the page differentiate as at DPQ. Draw BC parallel to the curve at D, and make PQ proportional to the quotient AB/AC. To derive a graph higher on the page integrate as at EFRS, where EF is in principle to be made proportional to the shaded area left of RS. Because this area lacks a natural boundary on one side integration is an inherently ambiguous process, from which definite values emerge only when we supply an agreed time-reference as described in the text. In integrating from the 'rate of increase' graph, the area HJK will be reckoned negative. With imperfect data integration can be

The most direct way of providing algebraically for the retardation of growth which real organisms display in later life is to introduce into the equations an arbitrary limit A, a maximum size which x may approach but can never surpass. An investigation by Reed (1920) is exceptional in the prominence which it gives to this idea of a limit. The subject was the elongation of shoots on mature apricot trees. The growth of a twig from a winter bud is a specialized process, involving the rapid expansion of preformed tissues and organs, and powered by reserve materials so that current photosynthetic output is not a limiting factor. Reed simplified matters still further by taking his observations rather infrequently so that any initial hesitation in the bursting of the bud would escape attention. He had therefore to deal only with measurements in which growth rate appeared to be a maximum at the start of the season, and thereafter declined continuously. Let us put:

$$\frac{\mathrm{d}x}{\mathrm{d}t} = k\,(A - x) \qquad (5)$$

(absolute growth rate proportional to the growth which remains to be accomplished), integration of which gives:

$$\log\,(A - x) = -kt + C \qquad (6)$$

where C is again the indeterminate constant of integration. The scale of time is here most conveniently standardized by making $x_0 = 0$, which brings us to:

$$x = A\,(1 - \mathrm{e}^{-kt}) \qquad (7)$$

In an experiment on the effect of pruning Reed found that k could be assigned the same value for pruned and unpruned trees alike, but that the pruning operation, performed in winter, more than doubled the value of A for the shoots growing out in the following spring and summer. In this situation the equations have a real function to perform, but it is important not to lose sight of the fact that their agreement with observation is largely a forced fit. The behaviour of a young shoot is

---

made reasonably accurate, but differentiation is much less reliable. These examples were constructed by drawing a free-hand shape for the second differential and integrating twice. The point of inflection of the sigmoid curve is indicated by the feathered arrow and is an almost imperceptible feature. With perfect data it could be located by differentiating twice to find the point H, but this is rarely a practicable mode of investigation. Inflection here being at one-third of the time-period shown, a Gompertz fit would be appropriate.

being accounted for in terms of a constant which is fundamentally a statement of the shoot's ultimate potentialities for growth in an idealized and infinitely distant future. In practice the constant A has to be made appreciably greater than the length actually attained, is given whatever value the observations require, and offers no explanation for anything. Reed's discovery was essentially that the shoot of a pruned tree is longer than the shoot of an unpruned tree by a fixed factor which is the same at every stage of growth. The equations reveal the simplicity of this result, but that appears to be the limit of their usefulness.

In attempting to represent the whole growth of an organism, from youth to old age, an obvious line of approach is to consolidate equation (1), which gives accelerating growth in early life, with equation (5), which records a decline and eventual cessation of growth activity. By writing:

$$\frac{\mathrm{d}x}{\mathrm{d}t} = kx\,(A - x) \qquad (8)$$

we obtain a smooth transition from the conditions of compound interest at the beginning to those of Reed's investigation at the end. When x is small $(A - x)$ will be virtually constant (substantially equal to A) and growth rate will increase with x. Towards the end x will be almost constant (nearly equal to A), and growth rate will diminish with the dwindling value of $(A - x)$. Integration from (8) gives:

$$\log x - \log\,(A - x) = kt + C \qquad (9)$$

where C is the constant of integration. It is convenient to fix the time-scale by specifying t_1, the time at which $x = \frac{1}{2}A$. This makes $C = -kt_1$ so that:

$$\frac{x}{A - x} = \mathrm{e}^{k(t - t_1)} \qquad (10)$$

This is the equation of a symmetrically sigmoid (S-shaped) curve; because of a chemical analogy it is generally known as the autocatalytic equation. It was transferred into biological usage directly from its chemical application, and has been a popular form of growth equation ever since it was recommended by Robertson (1908) and others.

A sigmoid curve (Fig. 22) has a point of inflection where its gradient is steepest. This is where the absolute rate of growth, which has so far been increasing, reaches its maximum value and begins to decline. At the point of inflection the growth curve is straight, though only momentarily. In the autocatalytic curve the point of inflection occurs when $x = \frac{1}{2}A/$, just half of the potential amount of growth having been accomplished. Many growth processes display a sufficient degree of symmetry about this midpoint to be reasonably well fitted by an autocatalytic equation with suitably chosen constants. It is important to notice, however, that the fixing of the point of inflection directly from observational data is not a practicable operation, the point being distinguished only by zero curvature in a long stretch of line where the curvature is very near to zero throughout. In deciding whether a set of observations can properly be treated as symmetrical one must in reality be guided by the existence of equal but opposite curvatures towards the ends of the graph. Despite this situation, so unfavourable to clear discrimination between the claims of competing forms of equation, it became clear at an early date that some growth processes are markedly unsymmetrical and reach their point of inflection well before half the final size has been reached. There have been several attempts to provide equations which would fit unsymmetrical growth data. Thus Gregory (1921) in a study of cucumber leaf growth obtained some data which could be fitted to his satisfaction by compound interest or autocatalytic expressions, but in particular conditions of temperature and high light intensity he found a need for unsymmetrical curves. His response (Gregory, 1928) was to superimpose upon a compound interest law the notion of a damaging time-factor. Excessively high temperature, for instance, may be conducive to rapid growth but may also do harm to the organism, and the amount of that harm may increase with the passage of time. Gregory's proposal therefore was to replace:

$$\frac{1}{x}\frac{dx}{dt} = k$$

(constant relative growth rate in compound interest) by:

$$\frac{1}{x}\frac{dx}{dt} = \frac{k}{t^n}$$

where n is a new quantity, which is to be supposed to increase rather steeply with temperature above a certain temperature threshold. The constant k also increases with temperature. The relative growth rate engineered in this way will have at low temperatures a moderate but sustained value, but at high temperatures a large value with rapid decline. This is an effective curve-fitting device which could be applied to other growth equations if required, but it is philosophically unattractive because no physical meaning can be attached to fractional powers of time. The obvious artificiality of Gregory's algebra sufficiently explains why it has not been generally adopted.

A much more elegant modification of compound interest was developed by Gompertz (1825), who was concerned with actuarial problems and worked out his argument in terms of 'a man's power to avoid death'. He supposed that power to diminish at a fixed percentage rate during every short period of time passed. The individual's prospects of survival are in other words to be made subject to a negative rate of compound interest. In order to use the Gompertz scheme as the foundation of a growth curve, we merely substitute relative growth rate for the power of survival. The Gompertz programme then uses the idea of continuous compound interest not once but twice over; the organism increases by compound interest at a percentage rate which diminishes by compound interest. Whereas Blackman had:

$$\frac{1}{x}\frac{dx}{dt} = k$$

$$x = x_0 e^{kt}$$

the fundamental assumption now becomes:

$$\frac{1}{x}\frac{dx}{dt} = k_0 e^{-bt}$$

where k_0 is an initial growth rate and the constant b fixes the speed at which the growth potential represented by k_0 is to be eroded by the passage of time. Integration and the rearrangement of constants will yield:

$$x = Ae^{-e^{(a-bt)}}$$

Except that we have adhered to our own previous usage in relation to A this is the standard Gompertz formula. It gives a markedly unsymmetrical sigmoid curve, which reaches its point of inflection when about 37% of total growth has been accomplished. Its application to growth studies has been discussed by Winsor (1932).

There is no limit to the number of equations which can be devised to fit growth data, and the literature offers examples of formulae which have not figured in the foregoing account. It would not be profitable to attempt to catalogue every variant. In the main they fall readily into two categories. Some authors have introduced modifications of a biologically intelligible kind into one or other of the equations we have already presented. If, for example, we substitute for the autocatalytic:

$$\frac{1}{x}\frac{dx}{dt} = k(A - x)$$

the new form:

$$\frac{1}{x}\frac{dx}{dt} = k(A - bx)$$

we thereby obtain a formulation, though not the one most commonly used, of the logistic equation, a different but still symmetrical sigmoid relationship. The logistic, though perfectly usable as a growth equation, has found its main application in population studies (Pearl & Reed, 1920). Other writers, particularly the earlier ones, resorted to the use of expressions which, however convenient as algebraic devices, could lay no claim to any measure of biological plausibility. The square root of a length, for instance, is a quantity totally void of biological meaning, yet there is no manipulative difficulty in so combining square roots as to generate sigmoid curves. The use of such equations has sometimes been advocated, but can command no support at the present day.

It is clear from general mathematical principles, and was sufficiently demonstrated by Gray (1928) in the present special connection, that the accuracy with which a 'theoretical' curve will fit a particular set of observations is not at all closely linked with the adoption of a particular form of equation. Thus for example:

$$kt = \log \frac{x(a - x_0)}{x_0(a - x)}$$

yields a sigmoid curve which inflects at $x = \frac{1}{2}a$, while:

$$kt = \log \frac{a - 3\sqrt{x_0}}{a - 3\sqrt{x}}$$

gives a sigmoid curve inflecting very near to $x = \frac{1}{3}a$, yet the two forms are so similar graphically that they would be virtually interchangeable if anybody chose to adopt them for growth analysis. That observed graph-points fall, for instance, upon an autocatalytic line, is therefore not in any sense to be regarded as evidence that the plant was growing in accordance with autocatalytic principles. Given the margins of observational error which prevail in this class of work, the points can quite certainly be fitted at least as well by a multitude of other equations, the number and structural diversity of which need be limited by nothing but our resources of mathematical skill and creative imagination. That an equation derived from some theoretical notion about the nature of growth should closely agree with observed growth rates is satisfying, but the satisfaction has to be tempered by the reflection that biologically meaningless algebra can be used with equal success. In reality the standard of precision which can be attained in the conditions of most growth investigations is not very high, and the uncertainties attaching to the location of the point of inflection are quite enough to eliminate any need for very precise adjustment of the form of the fitted curve. In the majority of cases one may proceed satisfactorily by using an autocatalytic equation where the data are roughly symmetrical, and a Gompertz equation where the point of inflection obviously comes before the middle.

An example of Gompertz fitting can be seen in the work of Amer & Williams (1957) on the growth of the leaf in *Pelargonium*. The leaf here continues growth for about eight weeks, but reaches its maximum growth rate after only one week, at which time the size of the leaf is very much less than half its final value. Such a pronounced asymmetry rules out any possibility of a worthwhile autocatalytic or logistic fit. Theoretically it should also rule out any prospect of a Gompertz fit, the asymmetry of the Gompertz curve being substantially less than that demanded by the material. Amer & Williams were nevertheless able to use the Gompertz (in the form: leaf area $= Ka^{b^t}$) with reasonable satisfaction. The fit was imperfect to an extent which is just noticeable in their published graphs, but was amply good enough to allow them to distinguish that the watering regime of the plants had a dramatic influence upon the values of K and a but very little influence upon the value of b.

We have so far dealt with the increase of a single measured quantity, such as a length or weight, considered in isolation as a selected index of the growth of the organism. When two or more increasing dimensions of the plant are kept simultaneously under observation, new considerations are brought into play. It will be necessary to study the situation which arises when two dimensions are concurrently measured and one is then used instead of a conventional time-base in preparing a graph showing the increase of the other. A typical example might show the breadth of a growing leaf plotted against the length of the same leaf. Any graph of this nature may be called a correlation curve.

It has been demonstrated in a large number of cases involving both plants and animals (Needham, 1942) that the doubly logarithmic plot (in our example the logarithm of the breadth against the logarithm of the length) approximates to a straight line. This signifies only that for a given percentage increase of one variable there is a proportionate percentage increase in the other. We would find such a relationship, for instance, if for every 5% increase in the length of a leaf there were a 3% increase in its breadth. This situation, usually known as an allometric relationship, implies a progressive change in shape except in the special case where the two variables increase by the same percentage so that the allometric line has unit slope. It must be noted that the maintenance of an allometric relationship, or of any other growth correlation, does not in principle depend on the maintenance of any particular rate of growth. In our leaf example it was not specified that a 5% increase in length should occur per day or indeed in any constant period of time, but merely that when it did occur it should be accompanied by a 3% increase in breadth.

An allometric representation is often a useful tool in studies of growth. To determine the slope of the allometric line is a convenient and effective test for changes of shape in the course of development. Thus, for example, Njoku (1956) in examining a sequence of leaves upon a shoot effected a considerable simplification when he resolved his data (Figs. 23, 24, see p. 34) into a set of allometric lines with unit slope, thereby demonstrating that the differences of shape with which he was concerned were established very early in development, and remained unchanged through the major period of leaf expansion. No other form of mathematical manipulation would have brought out this result so neatly. Another application can be seen (White, 1954, 1954a) in the study of developing leaves in *Phaseolus*. Here the lamina maintains a substantially unchanging shape (allometric constant for length and breadth almost unity) but the cross-sectional area of xylem in the petiole increases much more slowly than the area of the lamina (allometric constant for xylem area against lamina area only 0·61).

As a theoretical principle of growth physiology, however, there can be no doubt that the importance of the allometric formulation was greatly overestimated by many of the earlier writers. The practical implications of doubly logarithmic plotting were examined by Pratt (1941), who found by trial that quantities which are individually growing according to sigmoid curves tend very generally to yield correlation curves which are approximately allometric. It was afterwards proved (Dormer, 1965)

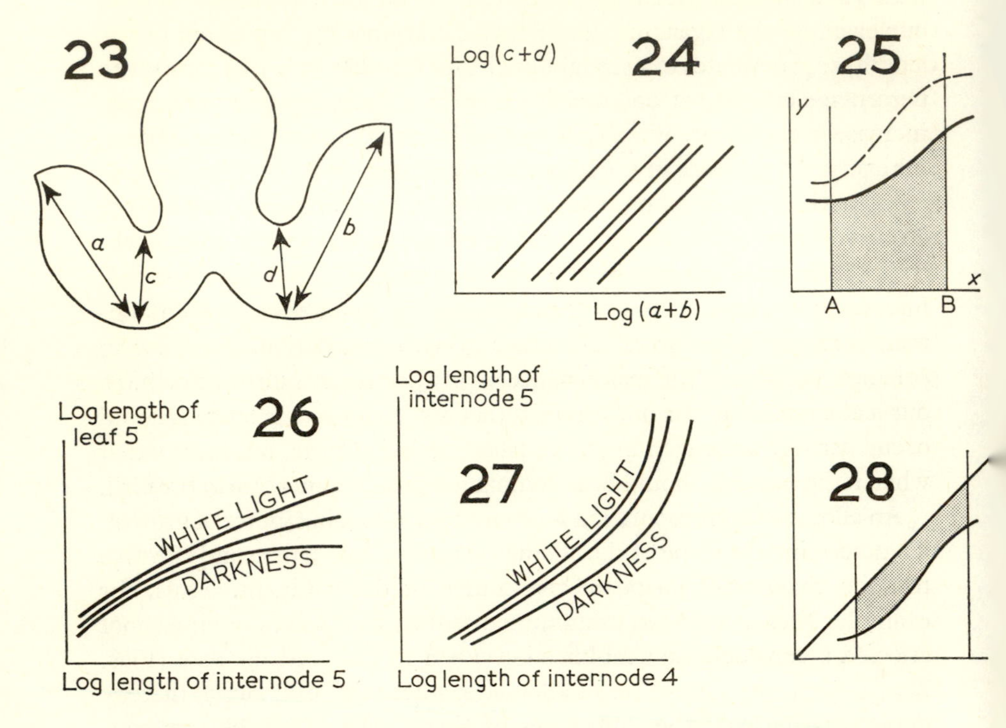

FIGURES 23–28. *Graphical treatment of growth correlations.*

23. System of measurement applied to lobed leaves by Njoku (1956), the quotient $(a+b)/(c+d)$ being taken as a 'shape index' which is larger as the lobing is more pronounced. **24.** Shape test by allometric plot. Each line represents development of one leaf. (Straightness of the lines was assumed as a legitimate simplification, the data being scarcely adequate for critical study of this point.) Successive leaves up a plumular shoot become more lobed, allometric lines consequently displaced to the right. That the differences in lobing are established very early in the growth of the individual leaf is demonstrated by the approximation of all these lines to unit slope (average slope for ten leaves was $1{\cdot}02 \pm 0{\cdot}04$). **25.** Definition of comparison integral (see text, p. 35). **26** & **27.** Correlations observed by Thomson & Miller (1961) in pea seedlings grown under differing conditions of illumination, the middle graph in each set being obtained in red light. In spacing and curvature these lines are typical of such investigations; with inadequate data and doubly logarithmic plotting an 'allometric relationship' (i.e. a straight-line fit) will often be reported, but these observations were too good to permit this. **28.** Definition of correlative plastochron (see text, p. 37).

that there is a special algebraic connection between the Gompertz equation and the allometric equation, such that an accidental coincidence in the values of two constants in the growth equations will automatically cause the appearance of an allometric correlation. It further appears (Dormer, 1965) that the 'allometric' nature of published observations has often been too uncritically accepted by the original author. When data are re-examined, some curvature of the claimed 'straight' line is often revealed. In the face of these developments it seems futile to follow the older authors in their search for a special physiological explanation for the 'allometric law of growth'.

A more profitable approach to the study of growth correlations can be founded upon the realization that the growth of a shoot involves a complex network of such correlations, and that a properly organized system of correlation curves would constitute a complete mathematical framework for investigations in developmental morphology. The general form of such a treatment is already clearly apparent, though there is still a great shortage of observational material. If we consider in the first instance the correlations existing in any one leaf and/or internode (for which purpose it is convenient to associate each leaf with the internode below), any given correlation curve, as for example a plot of petiole length against internode length, will have its own characteristic form, often but by no means universally approximating to the allometric, probably never exactly allometric if determined with sufficient accuracy. In general the form of a selected correlation will vary along the length of a shoot. Our proposed graph of petiole length against internode length will have different shapes for leaves 1, 2, 3 . . . of a seedling. There are enough examples on record to show that in a family of correlation curves obtained in this way the differences can usually be summarized by the method shown in Fig. 25. Here the limits A and B are arbitrarily fixed for a given set of comparisons, the continuous curved line is an observed correlation of two quantities x and y, and the shaded area may be called a comparison integral (Dormer, 1965). The broken curved line is the corresponding correlation for another leaf; it would give a higher value of the comparison integral. We may complete the working by showing in a final graph the way in which comparison integral varies with position along the stem. This procedure gives us a means of surveying the changes involved in a heteroblastic sequence, which may be defined as any progressive change in leaf-shape and associated characteristics from node to node along the length of a shoot.

As a comparison integral possesses many of the attributes of an average, and as it can reasonably be supposed that the form of a correlation curve will be relatively indifferent to moderate fluctuations in absolute growth rate, investigation by comparison integrals is likely to be a very sensitive test for revealing differences in the developmental history of successive leaves. Furthermore, if there is some systematic change in the physiological state of the growing shoot, if for instance some influential hormone should have a maximum or minimum concentration at some particular level in the stem, then this may affect simultaneously a high proportion of all the variables which are accessible to measurement. We must therefore anticipate that similar patterns of change may appear in sequences of comparison integrals based on different types of observation. This is illustrated in some work on *Vicia* seedlings (Dormer, 1950a, 1951; Dormer & Plack, 1951) where measurements of xylem differentiation and measurements of petiole length quite independently yielded comparison integrals with maxima in the fifth internode of the plumule.

The results of a physiological experiment can sometimes be represented better by the use of correlation curves than in any other way. Thomson & Miller (1961), who grew pea seedlings in darkness, in white light, and in red light, obtained a somewhat confused and inconclusive analysis when they plotted observations on the growth of their plants against a scale of days. The influence which different conditions of illumination exerted upon absolute rates of development was often very small; the rate of leaf production at the stem apex, for instance, was only slightly less in darkness than in full light. Nor did the results appear sufficiently consistent to permit the emergence of any simple generalization; darkness markedly retards the elongation of a young leaf, but paradoxically exerts a stimulating effect at a later stage so that the leaf grown in darkness achieves by a very late burst of activity a length surpassing that of its normal counterpart. When the same data are rearranged in the form of correlation curves a much more homogeneous set of relationships is revealed (Figs. 26, 27). If it were desired to examine in more detail the effects of coloured light upon the morphology of these plants, a promising line of development would be to set up agreed limits for the derivation of comparison integrals from curves of this nature, to expose plants on an equal-energy basis to light of different spectral bands, and so proceed to a chart of comparison integral against wavelength. The existing data point to the possibility that the behaviour of the plant in darkness might stand as a natural zero baseline for such charts.

A special status must be accorded to any correlation curve which is formed from corresponding variables of two successive leaves, as for example if we plot the length of the third leaf of a seedling against the length of the second. Curves of this category will very commonly show the kind of progressive alteration from node to node which has already been considered, but their special quality lies in their ability to express the interval by which the development of one leaf is delayed relative to that of its predecessor. In Fig. 28 suppose the curved line to represent a correlation of the kind proposed, and let the straight line above it be inclined at 45°, so as to represent the correlation which would exist between two leaves absolutely simultaneous in their development. Then the shaded area (similar to a comparison integral in its derivation) must be taken as a measure of the amount by which, in reality, the development of one leaf lags behind that of its predecessor. This question of the interval between successive leaves assumes considerable importance because of the widely felt need for a measure of physiological age. If we have two similar seedlings, each in process of unfolding its fourth leaf, it may be of little value in morphological work to give their ages in days. If one has been grown fast and the other slowly, a count of days will merely obscure the fact that both now appear to be at the same point of development. By reviving the old term 'plastochron' for the interval between one leaf and the next, we are enabled to express an important truth in the form that both plants have a physiological age of four plastochrons. Unfortunately it is not a simple matter to apply the plastochron concept to accurate quantitative work. It can be shown (Dormer, 1965) that there are two ways of defining the plastochron:

(*a*) The plastochron may be taken as the interval of time between corresponding stages of development in two consecutive leaves (or internodes).

(*b*) The plastochron may be taken as the difference in physiological age between two consecutive leaves (or internodes) of one and the same shoot, taken at the same instant of time.

The plastochron according to the first definition (which may be called the chronological plastochron) is simply a conversion factor, to be experimentally determined, between the leaf-production of the apex on the one hand, and clock-time on the other. Plastochron concepts developed in this way can be of great utility, but do not offer high standards of numerical accuracy. For one thing the conversion factor depends on the conditions of growth and will be subject, *inter alia*, to a high temperature coefficient. Even more seriously, it is impossible to find satisfactory procedures for combining observations from different specimens because individuals will differ by fractions of a plastochron, and the definition does not directly provide for the estimation of fractions.

The plastochron according to the second definition (which may be called the correlative plastochron) is not an interval of time but an interval of development, measured by the shaded area in Fig. 28. It can be shown (Dormer, 1965) that if the correlation curve in this diagram retained the same shape without alteration from node to node, then it would be possible to set up a fundamental scale of correlative plastochrons, with facilities for measuring fractional differences to any required standard of accuracy. This proposition, which is a strict algebraic relationship, is unfortunately of interest mainly in its negative form: as no real shoot is likely to satisfy the requirement of unchanging correlation, so the concept of a plastochron scale can never be more than a working approximation, often of great practical value, but inherently incapable of being improved beyond a certain limit of accuracy.

The utilization of a plastochron scale of age can be seen in the work of Erickson & Michelini (1957), who were interested in measurements of a destructive nature, in respect of which there could be no question of continuing observation upon a single specimen. It was necessary therefore to have means whereby measurements from different specimens could be made to contribute to a generalized curve. The problem essentially was the one of evaluating fractional plastochron differences in a scale based on the 'chronological' form of definition for the plastochron unit. Erickson & Michelini appear to have been at the outset quite unaware of the possibility of defining the plastochron on a correlative basis. They were influenced by Erickson's earlier experience with flower buds of *Lilium*, in which it had proved convenient to use a 'development index' which was the logarithm of the length of the bud. (On this basis, development index would be proportional to time if growth proceeded by a compound interest law.) It was proposed to extend this idea to the growth of leaves in *Xanthium* seedlings, the interval of time separating two successive leaves upon a shoot being distinguished as the plastochron unit. The criterion of leaf development was taken to be a length of 10 mm. Thus when the nth leaf of a seedling is exactly 10 mm long the age of that seedling is n plastochrons; similarly when the $(n + 1)^{\text{th}}$ leaf is just 10 mm long the age of the plant is $(n + 1)$ plastochrons. It remains to find a device for calculating the fractional plastochron-age which arises when the nth leaf exceeds 10 mm but the $(n + 1)^{\text{th}}$ leaf has not yet reached that length.

The treatment adopted by Erickson & Michelini assumes that leaves which are 10 mm long or not much more are growing by compound interest; if L_n is the length of the nth leaf the passage of time in constant conditions will be measured by $\log L_n$. As 10 mm is very much less than the mature size of the leaf this is a reasonable approximation, and at this standard length the graph of log leaf-length against time in days proved to be substantially straight for the first twenty leaves of a seedling. The further development of the method depended however on two additional assumptions, that the graphs for successive leaves would be parallel, and that they would be equally spaced. Observation (Fig. 29, see p. 40) showed that this degree of regularity was not maintained in practice. Some of the earlier leaves of the seedling (1 and 2, 3 and 4) showed imperfect pairing due to a subdecussate system of phyllotaxy; this is a complex disturbance, the nature of which will be examined in another chapter. Also the gradient of the lines, steep in the earlier leaves, had fallen considerably by the time the twentieth leaf came to expand. Erickson & Michelini, despite this difficulty, persevered with the development of their method. Upon their simple assumptions, when $L_n > 10 \text{ mm} > L_{n+1}$ we have:

$$(\text{plastochron age of plant}) = n + \frac{\log L_n - \log 10}{\log L_n - \log L_{n+1}}$$

where log 10 appears because a length of 10 mm has been adopted as the arbitrary plastochron standard. In order to facilitate the study of individual leaves the further step is taken of making:

$$(\text{plastochron index of leaf } n) = (\text{plastochron age of plant}) - n$$

which will give a negative index for any leaf shorter than 10 mm.

With all its faults this method offers over a considerable range of early leaf development a serviceable approximation. It means that measurements from separate leaves on separate plants can be combined into a single graph or calculation with an error of no more than an hour or two. The necessary estimates of plastochron-fractions are admittedly subject to certain residual errors, but they are much more accurate than anything that can be achieved by setting descriptive standards for intermediate stages in the maturation of a leaf. Erickson & Michelini examined, for instance, some earlier work on the leaf of *Pisum*, and showed that an interval which had been visually estimated as half a plastochron was really about 0·3. The method derives most of its value from the fact that it is essentially a system of interpolation. The two leaf lengths which are to be measured are always those which most closely

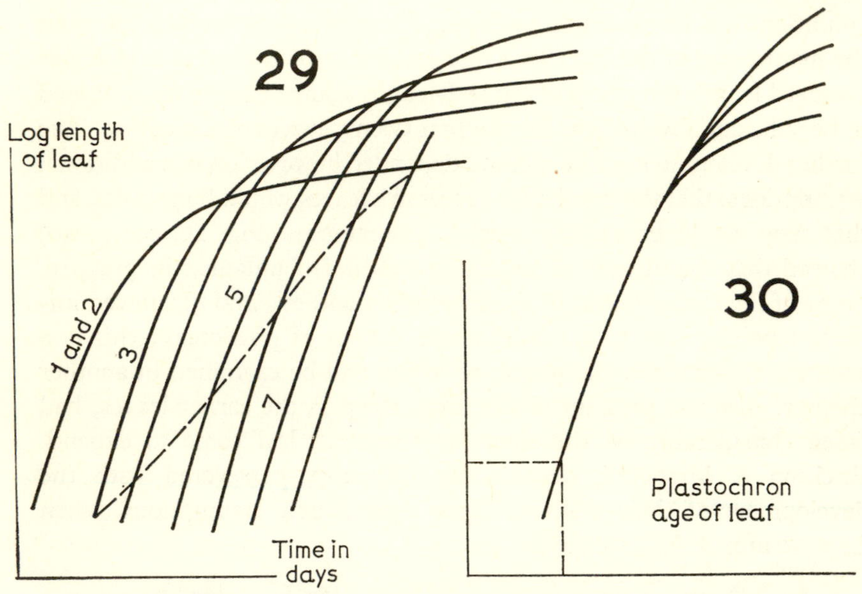

FIGURES 29 & 30. *Growth of* Xanthium *leaves* (*Erickson & Michelini* 1957).

29. Growth curves for elongation of successive leaves of a seedling, plotted logarithmically on a common time-base. The continuous lines are for leaves 1 to 8. As the difference between 1 and 2 was scarcely resolved in this investigation (the separation of the graphs being hardly more than the statistical uncertainties attaching to either) a single average curve is given for both. Whereas the later leaves (5 to 8 and beyond) show a fairly uniform spacing, the graphs for leaves 3 and 4 are visibly drawn together, with larger intervals between 2 and 3, 4 and 5. The broken line is the curve for leaf 17, transferred from its true position for comparison of gradients. A significant reduction in growth-rate sets in at about leaf 12. This flattening of the curves sets a limit to the applicability of the simple logarithmic calculation of plastochron age in one direction, as the irregular timing of the first four leaves sets a limit the other way. **30.** Part of the same data replotted on a basis of plastochron age of leaf. A leaf length of 10 mm, giving the zero of the plastochrone scale, is indicated by the broken lines. The form of graph is common to various leaves up to a length of about 50 mm. Later stages diverge to give characteristic mature lengths which increase progressively up to about leaf 9 or 10, and then diminish again.

bracket the chosen standard of 10 mm, and whatever the plastochron age to be determined its integral part is necessarily correct. Only the fractional part of the estimate, with an average value of half a plastochron, is vulnerable to errors arising from phyllotactic irregularities etc. Reference to Fig. 30, which shows the growth of several leaves brought to a common time-base, will make clear the principle that differences in the later life of the leaves do not affect the reliability of measurements near the 10 mm standard. A point of some theoretical interest emerges from these discussions in that it is possible to recognize a 'relative plastochron rate of elongation' given by:

$$\frac{\Delta\,(\log \mathrm{L})}{\Delta\, n}\ (t \text{ constant})$$

Except that n is discontinuous, this is an exact counterpart of relative growth rate:

$$\frac{1}{L}\frac{\mathrm{d}L}{\mathrm{d}t} = \frac{\mathrm{d}\,(\log L)}{\mathrm{d}t}\ (n \text{ constant})$$

One of these expressions shows how leaf length increases with time, the other how it increases with distance from the apex. In a shoot which satisfied the Erickson & Michelini conditions (as the *Xanthium* plumule does not) the two would be interchangeable, and measurements of successive leaves backwards from the apex could be plotted directly to yield a growth curve scaled in plastochrons. This would be a scale of correlative plastochrons in the sense of Dormer (1965); the two cases are completely congruent both as to the result which would be obtained and the conditions which would have to be satisfied.

The plastochron scale established for the *Xanthium* shoot was utilized by Maksymowych & Erickson (1960) in their study of the histogenesis of the lamina. Some of the principal events in the growth of the leaf are listed in Table 2 against a scale of leaf plastochron index. The ability to draw up this kind of chronology represents a great advance upon purely descriptive treatments and upon cruder systems which merely state the size of the primordium at each stage. Some remarkable features appear in this record.

Out of a total immature life of more than ten plastochrons no less than eight are passed in a substantially unventilated condition, while between the early mitoses which bring the lamina to a thickness of six cells and the later mitoses which add further layers there is an interval of no less

TABLE 2

Time-table of significant events in the development of *Xanthium* leaf, using plastochron scale for the physiological age of the leaf. From Maksymowych & Erickson (1960)

Plastochron age of leaf		Histological events
Before	−6·0	Leaf-primordium already visible
	−5·0	First signs of marginal meristematic activity
	−3·8	Procambium formed
	−2·6	Lamina six cells thick
	+1·0	Mitoses to give additional layers
	+1·7	Last mitosis in lamina
	+2·0	Rapid increase in thickness, intercellular spaces appear
	+4·0	Thickening ceases, leaf mature

than 3·6 plastochrons, more than a third of the whole period surveyed. Without a sufficient system of chronology it would have been impossible to gain any clear impression of these relationships. The presentation in the table remains, however, at an elementary level from the mathematical point of view, because the development of the lamina is dealt with in terms of qualitatively distinct events. The attempt to consider development as a continuous process will inevitably raise once more the question of 'plastochron rate', of quantities analogous to growth rates but based on a plastochron scale.

The nature of this development can be seen in a study by Maksymowych (1962) of the area growth of the *Xanthium* leaf. The technique employed was the simple one of marking a leaf with indian ink and observing the displacement of the marks. All the refinement of the investigation came in the analysis of results. What Maksymowych measured was the distance X between a given mark along the midrib-petiole axis and the tip of the leaf, the latter serving as origin of the co-ordinate system. Because the tissue is expanding, the ink-mark is moving away from the tip of the leaf at a velocity (not constant) of $\mathrm{d}X/\mathrm{d}t$. As development was to be studied on a plastochron scale, interest centred on the analogous plastochron rate, which may be written as $\mathrm{d}X/\mathrm{d}P$ where

P is the leaf plastochron index (or plastochron age of the leaf). The velocity of the ink-mark is obviously a summation of local growth rates, every piece of tissue between the mark and the leaf-tip making its own contribution. If we consider the expansion of the tissue in any thin transverse slice of the leaf this local rate of expansion will appear as a difference between the velocities of the apical and basal faces of the slice respectively. The 'relative elemental growth rate', the factor by which the slice (or element) increases per plastochron, is therefore to be obtained by differentiating a second time, this time with respect to X; relative elemental growth rate is:

$$\frac{d}{dX} \left[\frac{dX}{dP}\right]$$

Maksymowych was able to calculate this quantity from his observations by reputable processes of numerical analysis, and so to obtain a comprehensive picture of the distribution of growth in leaves of various ages (Fig. 31, see p. 44). In the young leaf the rate of expansion is practically uniform, except at the extreme tip, but as the leaf ages the rate falls to a very low level progressively from the tip downwards, until in the last stage it is only the petiole which continues to elongate.

It has been possible to bring the discussion to this point by treating growth as a simple linear movement; we have considered the amount of growth which may occur, but have not been specifically concerned with its direction. Growth, however, being of the nature of a velocity, is evidently a vectorial quantity, and this aspect of the matter has to be taken into account as soon as any piece of tissue begins to grow at different rates in different directions. Direct three-dimensional observation of movements in solid tissue masses has not yet been practised to any significant extent, but we cannot set aside the corresponding two-dimensional problem as it may occur, for instance, in the extension of a leaf-lamina. Suppose a leaf to grow so that its length and breadth are related by an allometric coefficient differing from unity; then any marked point in the lamina will be describing a curved trajectory relative to the midrib. Without attempting a complete analysis, some attention must be given to the botanical implications of this more complex type of movement. The purely observational part of an investigation will rarely present much difficulty. Either by making use of recognizable points in the venation, or by applying suitable ink-marks, one may observe the growth of the leaf over a period and prepare a map

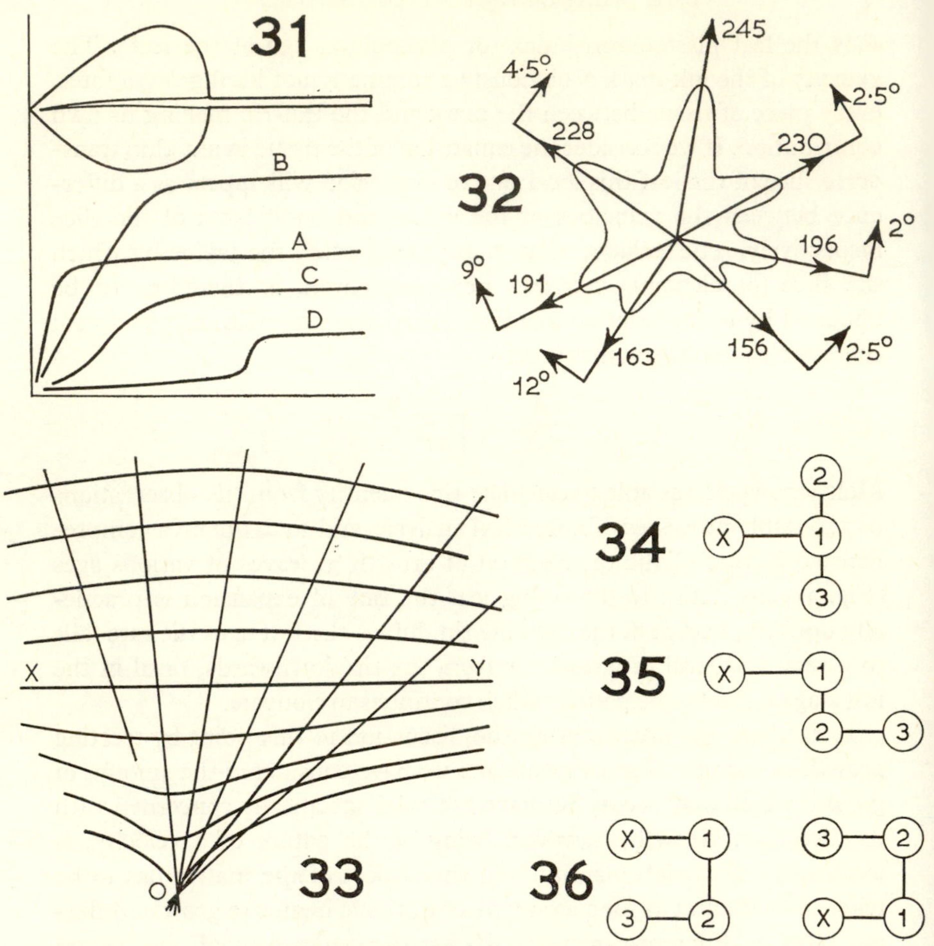

FIGURES 31–36. *Growth in leaves and inflorescences.*

31. Results (not to scale) of estimation of relative elemental growth-rates in *Xanthium* leaf by Maksymowych (1962). ABCD are successive distributions of growth activity along the length of the leaf, as indicated by conventional shape above. **32.** Changes in shape of maturing *Tropaeolum* leaf (Smirnow & Zhelochovtsev, 1931); outline shown is that of a very young leaf, the seven principal veins of which have been extended and marked with numbers denoting radial and angular displacements which are to take place before maturity. Radial extensions will be such as to accentuate the existing prominence of the upper lobes, but the asymmetry of shape will be reduced by large angular displacements at lower left. **33.** Principle of conchoidal distortion of a growing margin; family of conchoids related to a point O and a straight line XY, all radii from O being equally divided by successive curves (see text, p. 45). **34–36.** Three types of lateral cyme in *Ranunculus* (Cunnell, 1961); each having three successive flowers springing from the parent axis X. Each type exists in two mirror-image versions, as represented here in Fig. 36 only (see text, p. 53).

showing, at each selected stage of development, and for each of the chosen points, both the rate and the direction of the movement of that point with reference to some arbitrary coordinate framework. The necessary comparisons are best made photographically, with all stages of development enlarged to a standard size, and the choice of leaf will often be governed by the desire to simplify the task of plotting the movements. Smirnov & Zhelochovtsev (1931) chose the leaf of *Tropaeolum*, which presents a rounded outline at maturity though prominently lobed when young, and which lends itself to the application of a polar coordinate net, with seven principal veins radiating from the pole. The Russian workers developed several ways of dealing with the change in shape of the leaf during the course of growth. At the centre of Fig. 32 is the outline of a very young leaf, at a stage which can only be revealed by destructive dissection of the apical bud. The asymmetry is natural, though not here represented to scale. The seven principal veins have been extended beyond the tips of the lobes, and to each two numbers are attached, one being a comparative value for the extension which is to take place along the line of the vein before maturity, the other the angular displacement which the vein will undergo towards the tip, the angular displacement of which is accordingly zero. These figures show that the angular displacements are such as to reduce the original lateral asymmetry, but that the radial elongations are so distributed as to enhance the already conspicuous supremacy of the apical lobe over the lateral and basal ones. Evidently conditions in the regions between the lobes will vary according to the angular displacements; between the two basal lobes the tissue will be required to grow tangentially in a manner not demanded of the areas flanking the terminal lobe.

It can readily be shown that in the growth of a flat sheet the direction of growth is as important as its magnitude. In Fig. 33 let growth proceed radially from the point O, and let the straight line XY represent the edge of a sheet at some particular instant. Mark off each radius in small uniform intervals, working in both directions from the intersection of the radius with XY. Draw smooth curves through corresponding graduations on all the radii. These curves are conchoids, those below XY, which are concave upwards, representing past stages in the history of the edge of the sheet, while the conchoids above XY, which are convex upwards, are shapes to be attained in the future. This 'conchoid principle' can be generalized in several ways. It is not necessary that XY should be straight, or even a conchoid, for uniform growth along the radii to bring about changes in shape. Nor will the principle lose all its force if the 'radii' become inconstant in their divergence. In any circumstances where equal absolute amounts of growth are directed along lines which are unequally inclined to the margin, there will be a change of shape.

By photographic methods Smirnov & Zhelochovtsev were able to plot upon an outline of the young leaf a complete distribution of the radial component of the growth remaining to be performed. For this purpose the tangential part of the movement was simply discarded. The 'isolines' of growth which they obtained were in the central part of the leaf very nearly circular, much more circular in fact than is the natural contour of the leaf at any stage of its development. Towards the margin the lines rather imperfectly follow the shape of the leaf; along those intermediate radii upon which the leaf margin comes nearest to the centre of the leaf the 'isolines' are somewhat closer together, implying a locally increased relative elemental growth rate. Modern computer technology provides us with greatly increased facilities for carrying out the analysis of data in investigations of this kind. It is of course a purely mathematical problem to estimate from the observed movements of selected points the elemental growth rates (in various directions, and at successive instants of time) of the tissues between those points. To have this manipulation reduced to a speedy and automatic routine is certainly desirable, but the improvement will be one in the precise specification of botanical problems, to the solution of which the mathematical developments will not directly contribute anything at all. Erickson (1966) has given a treatment based on the mathematical concept of divergence of vectors. This has the effect of generalizing the idea of relative elemental growth rate from the simple form used by Maksymowych: whereas Maksymowych considered merely a transverse slice of leaf tissue and obtained a longitudinal growth rate which was simply the algebraic difference between the speeds of the two boundaries of the slice, the computer programme used by Erickson dealt with growth outwards in all directions from each point upon the lamina. The computer gave fundamentally two types of print-out. In the first place it drew up a contour map of the leaf showing the distribution of the production of new area (i.e. local values for rate of increase of area per unit of existing area). Secondly it could present, in several different forms at choice, an account of the isotropy or anisotropy of the behaviour of any piece of tissue; in other words, it could indicate whether a small circle was tending to grow into a larger circle or into some deformed figure such as an ellipse. This is a complete analysis. So far as the facts of growth distribution are concerned, and subject to any deficiencies which may exist in a particular set of observations, it leaves nothing undiscovered. It does not, however, provide any explanation for the phenomena.

In designing equations for the representation of growth it is necessary to provide for the fact that most organs do not grow indefinitely but have a rather definite limit to their growth potential. It is sometimes convenient to write this consideration directly into the algebra (as is done in the autocatalytic equation with the introduction of the constant A) without attempting to identify the source of the limitation. The investigation of the way in which growth potential is distributed between different parts of a shoot, as for example between successive leaves upon a stem, is however a necessary further stage in our enquiry. Work in this direction begins almost at once to encroach upon the domains of later chapters, and at present we need do no more than is required to establish a continuity of ideas between simple growth measurement and the more complex problems which are to follow.

Many attempts have been made to establish relationships between the size or condition of a stem apex and the growth potential of the leaves produced from it. The simplest problem of this kind is perhaps that which arises in linear-leaved monocotyledons where there is really only the width of the leaf to be considered. Abbe, Randolph & Einset (1941) dealt with maize, using leaves 6 to 12 of the plumular series. A measurement of leaf width is here without meaning by itself, because the width increases with time, and the study was concerned with leaf width for a given length. A doubly logarithmic plot of leaf width against leaf length yielded slightly sigmoid curves which could practically be treated as parallel and uniformly spaced. Within the chosen range it appears that the width of leaf for any chosen length is a quantity which increases from one leaf to the next by about 9%. Within the same range the circumference of the meristematic dome at the apex of the stem was found to increase by about 4% per plastochron. Rather surprisingly, the authors concluded that these figures could be considered to be in agreement. Difficult though it might be, however, to establish the reality of a 5% discrepancy on a single comparison between one leaf and the next, there seems to be no doubt of its existence in the conditions of this investigation, where rates of change were compounded through six successive internodes. That the increase in the width of leaf can be regarded as a simple consequence of the increase in stem diameter appears rather unlikely.

Differences in the growth potential of leaves can be displayed not only as differences in size but also as differences in morphological elaboration (differences in number of leaflets, etc.). It was upon this level that White (1968) was able to relate the growth of the leaf in *Marsilea* to the size of the apical cell of the stem. In this study the area of the apical cell (which means the area in a median longitudinal section) was determined for shoots which had reached various stages in the heteroblastic sequence from the simple leaf of the sporeling to the four-leaflet pattern of the adult plant. There is an approximately linear relationship between leaflet number and apical cell area. By varying the cultural conditions it is possible to change the size of the apical cell, which becomes very much larger if the plant is given glucose, while a reduction in size can be brought about by administration of thiouracil, an inhibitor of protein synthesis. It was strikingly clear in White's investigation that these artificially induced changes in the size of the apical cell do not result in the changes of leaf-shape which one might expect. Supplying glucose does not merely increase the size of the apical cell; it also raises in similar proportion the threshold of size which that cell has to attain in order to permit the production of leaves of any specified level of complexity. In other words the enlargement of the apical cell which is provoked by glucose turns out to be irrelevant to the heteroblastic sequence. The cell is larger at all stages of development, but the stages are still in their normal relationships to each other. The law connecting cell-size with leaf-shape has been transferred to higher standards of cell-size, but otherwise operates unchanged. This observation might well be taken to indicate that the size of the apical cell did not directly influence the state of the leaf at all, but was itself determined by an interaction between environmental factors (such as supply of glucose) and an unknown system of heteroblastic control.

All such studies are essentially concerned with what is called in common language the vigour of the shoot. This ill-defined but still useful concept must be supposed to embrace both the regenerative capacity of the preceding chapter and the growth potential of our present discussion. Increased vigour may lead to increased growth of existing organs or to the initiation of larger numbers of new ones, and there is no general method for predicting which response will predominate in any particular circumstances.

The work of McCully & Dale (1961) on the initiation of leaves in *Hippuris* is a striking demonstration that the vigour of an apex may have an effect upon the numbers of leaves which it produces. The phyllotaxy here is absolutely verticillate; that is to say, there appears to be no lingering remnant of any spiral pattern, and all leaves of a whorl arise in a single transverse layer of cells across the stem (though not always quite at the same time). The position is complicated by the existence of both aerial and submersed shoots; the submersed shoot has a taller apical dome, for a given diameter, than the aerial one, or, which is another way of expressing the same thing, the submersed leaf is relatively slower to project above the surrounding stem surface than is the aerial leaf. Although measurements upon the two types of shoot must necessarily be kept separate, it does not appear that they differ in any respect which materially affects the first appearance of leaves. Leaf initiation begins in the hypodermal layer of the stem at a level where the stem surface is already cylindrical rather than flat, and McCully & Dale show very clearly that the number of leaves in a whorl is related, more than anything else, to the count of hypodermal cells round the circumference of the stem. The characteristic stage is that at which in transverse section through the hypodermal leaf-primordium patches, each of these consists of three cells; actually there are seven altogether, two on a plane above these central three, and two on a plane below. The number of cells in the hypodermal layer between the adjoining primordia of a whorl averages very regularly between 2·5 and 4·5, and is about the same in aquatic and aerial shoots. The number of leaves in a whorl varies from three to at least thirteen, does not usually remain constant along a shoot for more than two or three nodes together, and may reasonably be regarded as a function of stem diameter. While this is an interesting situation, it has to be recognized that the relationship discovered owes its existence to a type of phyllotaxy which is decidedly uncommon among the general mass of vascular plants. That the system of leaf-determination which operates in *Hippuris* should be equally influential in any large number of other plants does not appear very likely.

In some cases it is impossible to base a realistic assessment of growth potential on any of the criteria so far introduced, namely the size, number, or complexity of organization, of the parts produced. Some developmental processes are of such a nature that their characteristic result appears only upon completion of a full sequence of stages. If, for example, a flower is capable of carrying the growth of its parts through only 20% of the normal programme, the reproductive failure of that flower is total, and its performance, from most points of view, might as well be recorded as zero. Comparable examples can be selected from vegetative morphology. In respect of phenomena with these characteristics there can be very little interest in ascertaining the average amount of growth in a set of specimens. Attention turns rather to the proportion of specimens which complete a particular developmental task and the proportion which fail to do so. That the average development attained by flowers in a particular position was 90% of normal might mean no functional flowers at all or nine good flowers for every ten specimens, or any intermediate proportion; what is needed is a statement of the probability that a bud at the specified node shall reach an agreed standard of maturity.

A treatment in terms of probability is appropriate in the examination of branching systems. Very often the majority of the stem apices in a shoot system will be dormant axillary buds. The selection of particular shoots for further growth is a major factor in fixing the characteristic habit of the species. The work of Cunnell (1961) on the inflorescence of *Ranunculus bulbosus* has the advantage of dealing with a fairly simple system in which there is a sufficient range of individual variation to support some investigation of the quantitative laws involved. The plant has usually a single inflorescence. This is terminal in position, and has always a terminal flower, the axis of which bears any number of bracts from one to four. Any lateral flowers are borne in few-flowered cymes in the axils of these bracts. Within this scheme the options open to the plant at any particular node are extremely limited. It can produce a lateral flower or refrain from doing so, and at certain points it has also a choice between right- and left-handed configurations. These simple elements of choice, however, are enough to generate a large array of permutations. It appears from Cunnell's survey that, of the inflorescence types which his description enables us to recognize, the majority will have frequencies of occurrence of less than 1%. His sample of 1500 specimens was therefore too small, probably by a factor of several hundreds, to permit an elucidation of all the interactions between the different parts. It is easy, however, to summarize the main quantitative trends in the organization of the inflorescence as its complexity increases. In Table 3 the specimens are graded according to the total output of flowers and the number of bracts. Evidently the introduction of addi-

TABLE 3

Relationship between flower-output and number of main-axis bracts in inflorescences of *Ranunculus bulbosus*. Sample contained 313 specimens with 4 flowers: 3 bracts, and so on (observations of Cunnell, 1961)

Number of flowers	Numbers of bracts			
	1	2	3	4
1	10	46	37	3
2	7	270	103	4
3	2	207	164	16
4	0	110	313	17
5	0	9	48	15
6	0	2	41	18
7	0	3	26	13
8	0	0	1	14
9	0	0	0	1
Average numbers of flowers	1·58	2·67	3·63	5·17

tional bracts is accompanied not only by an increase in the average number of flowers (at foot of table) but by a broadening of the range of variation. In considering the correlation between number of bracts and number of flowers Cunnell rather surprisingly treated the latter as the independent variable. He obtained a graph which would point to the addition of about 3·8 flowers for each extra bract but that is upon the basis that the addition of bracts is caused by the increase in flower-production, which runs contrary to the order of development of the parts. An effect cannot very well precede its cause, and this line of thought may lead us to prefer the significantly lower flower-to-bract ratio, about 1·2, obtained from the averages at the foot of Table 3. Cunnell was averaging bract numbers in rows, we are averaging flower numbers in columns; it is a question, in statistical language, of choosing the relevant regression coefficient of the pair.

The analysis can be carried a stage further by ascertaining how, in inflorescences with different bract numbers, the total complement of lateral flowers (it would be pointless to count the terminal one, which is always present) is divided between the axillary cymes. In Table 4 the

TABLE 4

Division of the total complement of lateral flowers between successive lateral cymes in the inflorescence of *Ranunculus bulbosus*. Plants with different numbers of main-axis bracts kept separate, lateral flowers summing to 100% in each group (observations of Cunnell, 1961)

	Successive lateral cymes			
Number of bracts	1st	2nd	3rd	4th
1	100	—	—	—
2	87	13	—	—
3	65	35	0	—
4	51	39	10	—

entries are percentages, summing to 100 in each row, zero representing a small fraction. It can be seen that the addition of the upper bracts leads to a progressively more even distribution of lateral flowers among the axillary cymes of the lower bracts. Although the scope of the table is limited by the potentialities of this particular species, it is easy to see how matters could be expected to develop in other examples where more than four bracts could occur. The introduction of a fifth bract would certainly diminish still further the percentage contribution of the first lateral cyme and increase that of the third. Although the importance of the second cyme has been increasing up to the four-bracted condition it does not look as though that trend could continue much further. On the contrary, as the inflorescence became longer it would seem inevitable that each node in turn should reach a turning-point and thereafter diminish in relative importance (not, of course, in absolute flower-output). We may find in some such scheme the germ of a more general mathematical treatment for branching systems, and in this respect it is encouraging to find that the shape of a bract bears a relationship to its position. Not only are lower bracts more leaf-like (in the shape of the lamina, sometimes also in the possession of a distinct petiole) than upper bracts, but they are more leaf-like as more bract nodes separate them from the terminal flower. The first bract of a four-bract inflorescence is more leaf-like than the first bract of a two-bract inflorescence, and so on. To take a step which Cunnell did not, we may couple the question of bract shape with the structure of Table 4 and tentatively suggest the existence of laws in such form as: 'a bract can develop a petiole only when the inflorescence apex retains sufficient vigour to ensure that at least 35% of the lateral flowers of the inflorescence shall be carried at nodes above that of the bract in question.' This restricts the development of a petiole to the lowest bracts of three-bract and four-bract inflorescences (which is in accordance with observation) and is a credible form of relationship because it links the shape of the leaf to the physiological state of the stem apex which produces it. The idea will readily admit of further refinement.

A more detailed analysis of the behaviour of the inflorescence can be attempted by classifying the different patterns of lateral cyme. In Figs. 34–36 we have plan views (from above) of three types of axillary three-flowered cluster. For each of these diagrams it is possible to design an opposite-handed counterpart by putting the second flower on the other side of the first. Only in Fig. 36 have both the opposites been shown. Our three basic designs, however, differ from one another in other respects than handedness. Similar considerations would apply to more complex cymose constructions. So far as handedness is concerned, Cunnell's observations, though insufficient in number, tend to show that right and left counterparts are equally frequent, though probably not randomly distributed (there is a suspicion that a cyme tends to be followed by another of the same handedness). The basic patterns, however, are by no means equally frequent. Cunnell observed the structures of Figs, 34, 35, and 36 in 68, 35, and 3 instances respectively. The study of the system here changes its character, the main interest being no longer in the assessment of vigour but in the symmetry relationships which emerge. We may reserve the further discussion of these matters until Chapter Five, noting only the importance of distinguishing differences of handedness from more fundamental differences of construction.

Chapter Three

Cell-enlargement and Cell-division

The growth of any organ may be regarded, if we so desire, as the resultant effect of two processes, the increase in number of cells, and the increase in size of the individual cell. It is a commonplace observation that these two processes tend to be partially separated in time; in general very young organs are characterized by active mitosis without much cell-expansion, while tissues which have almost completed their growth inevitably pass through a final phase of pure expansion uncomplicated by the appearance of new cells. What we know of the physiological control of growth also tends to emphasize the relative independence of the two components. It is easy to find experimental treatments which will dramatically change cell-sizes without making much difference to cell-numbers and vice versa. The opportunities of growth analysis which are opened up by these considerations are undeniably attractive, but it is impossible to be content with a treatment of growth in which the size of an organ is regarded as a mere consequence of the size and number of its constituent cells. This is only one side of the question, and in order to attain a more balanced view it is necessary to recognize that an organ may have systems of growth regulation which transcend the cellular level of decision-making. Studies of organ-growth, in which cell-walls may be valued primarily as convenient growth markers, as a more refined substitute for the ink-spots of the early physiologists, afford a very necessary corrective to the somewhat over-simplified picture which tends to result from excessive concentration on the behaviour of the single cell.

Attempts to determine the relative importance of cell-division and cell-enlargement in the growth of any organ have more often been conducted by measurement of cells than by counting. In some respects, however, there may be advantages in giving greater prominence to the estimation of cell-numbers. Particularly in the growth of leaves, where there is commonly a high degree of standardization in the size attained at maturity, it is natural to enquire whether there may not be some principle which directly regulates the number of cells in the leaf, more or less independently of any factors which may influence their expansion. Some valuable methods for the quantitative study of the leaf lamina have been developed in pharmacognosy, where the identification of fragments of leaf material assumes considerable importance in connection with the purity control of drugs. Several of the recognized criteria for this kind of identification are of the nature of ratios between different types of constituent cell in the leaf; one may be able, for instance, to distinguish between two species by estimating the average number of palisade cells underlying each cell of the upper epidermis. Gupta (1961) has combined the kind of tissue-sampling procedure which has long been customary in pharmaceutical work with direct measurements of the area of the leaves from which the samples are taken. In this way it is possible to choose any particular structural feature (as for example a specified type of epidermal hair) and to estimate how many times that feature is repeated in a given leaf. Gupta took five species of Solanaceae, and determined for several leaves of each the total number of repetitions of three different morphological features: (*a*) the vein islet, the smallest closed mesh of the leaf venation, (*b*) the veinlet termination, which is any vascular strand which ends blindly within a vein-islet, and (*c*) the stoma. By calculating a little further than was done in the original we can obtain the results given in Table 5. None of the structures listed here is a cell, but this hardly detracts from the interest of the observations because it is very generally

TABLE 5
Absolute numbers of parts in leaves of Solanaceae (calculated from Gupta, 1961)

	Absolute vein-islet number (V_I)	Absolute veinlet termination number (V_T)	Absolute stomatal number (S)	Ratios	
				$\frac{V_T}{V_I}$	$\frac{S}{V_I}$
Nicotiana tabacum	214,400	135,100	3,540,000	0·63	16
N. plumbaginifolia	20,200	18,800	541,000	0·93	26
Solanum nigrum	11,100	11,000	281,000	0·99	25
S. ferox	72,500	36,300	2,221,000	0·50	30
Withania somniferum	29,700	22,300	286,000	0·75	9

found in this class of work that the constituents of the leaf in any species tend to occur in fixed numerical proportions. Estimates for instance of the number of palisade cells, or even of the number of cells of all kinds, would give different absolute values but would be unlikely to follow a different pattern of behaviour.

The very large interspecific differences appearing among the absolute numbers in Table 5 are mainly due to differences in mature leaf-size. Absolute vein-islet number varies in a ratio of almost 20:1, but variation among these species in the number of vein-islets per unit area would hardly exceed 3:1. Observations were taken upon sequences of leaves along a shoot, including immature leaves near the apex. From these measurements it appeared to be a general rule that every young leaf passes through a stage in which, owing to the smallness of its area, all the characteristic features are densely crowded together, even though they may not yet exist in their full number. A leaf in this phase of development may for instance possess only 90% of the stomata which it is to have at maturity, but the number of stomata per unit area of its surface will be much greater than in a mature leaf. In all these species the absolute numbers with which we are concerned are reached some time before the expansion of the lamina is completed, and there must be a strong presumption that the growth of the leaf closes with an extended period of almost pure cell-expansion. This consideration tends to emphasize the importance of absolute numbers, because it means that a significant fraction of the total area growth takes place in conditions where size is governed by numbers rather than the other way round. It suggests also that ecological factors which influence leaf-area may not necessarily make the same impact upon absolute numbers of parts.

The ratios given in the last two columns of Table 5 were not calculated by Gupta and are not among those in general use as criteria for identification of specimens. They illustrate however the general principle that the tissue-elements of the lamina occur in proportions which characterize the species. It is not likely that any such ratio can be entirely without value as a taxonomic character. The constancy of the proportions for any species is another aspect of the absolute number concept; two leaves which differ only in their degree of cell-expansion will have different numbers of parts per unit area but will not differ, even locally, in any of their ratios.

The measurement of the sizes of mature cells was practised at a very early date, but seems not at first to have been related to any clear idea of growth analysis. Simple curiosity about the taxonomic situation (does an elephant have larger cells than a mouse? etc.) was the driving force behind many investigations which collectively failed to produce any consistent trend. For example Tenopyr (1918) made an excellent series of comparisons between plants with broad and narrow leaves. Sometimes the broad and narrow types were allied species, sometimes they were only of varietal rank, and sometimes they were merely heterophyllous modifications upon the same individual. It was found throughout that the differences in leaf-shape were reflected very faintly, if at all, in the shapes of the cells. In order to extract significant information about development from observations on mature cells it is necessary to adopt more complicated procedures, as in a study by Sinnott (1930) which not only incorporated detailed comparisons between the different tissues of an organ but considered simultaneously the response of the cells in both longitudinal and transverse directions. The material used was the petiole of *Acer*, and the differences in size arose as part of the system of dorsiventral symmetry which characterizes horizontal twigs. Complete genetic homogeneity was assured by taking all the material from a single tree. The range of variation was very great, the volume of an individual petiole being from 5 to 684 mm^3, and the investigation proceeded mainly by the use of correlation coefficients. So far as the parenchymatous tissues (epidermis, cortex, and pith) are concerned, the principal relationships are sufficiently shown in Table 6, the figures of which have been extracted from Sinnott's more extensive table.

TABLE 6

Correlation coefficients between measurements on *Acer* petioles showing non-genetic variation in size (extracted from Sinnott, 1930)

	Tissue examined			
	Epidermis	Hypodermis	Sub-hypodermis	Pith
CORRELATIONS BETWEEN:				
Cell-diameter and petiole diameter	+0·011	+0·633	+0·637	+0·836
Cell-length and petiole length	+0·288	+0·400	+0·600	+0·738
Cell-length and cell-diameter	−0·341	−0·039	+0·487	+0·338

The outstanding discovery here is that the cells of the parenchymatous tissues show increasing size-correlation with the organ in which they are situated as they are further removed from the surface. Even in the pith this correlation is not perfect. A longer and thicker petiole will have greater numbers of pith cells, both in length and in cross-section, but the increase in cell-numbers will not be in proportion to the increase in organ size. It remains true that increasing size of petiole is reflected in the pith mainly by increase in cell-size, but in the epidermis mainly by increase in cell-number, while tissues in intermediate positions display intermediate patterns of adjustment. Particularly striking is the indication that epidermal cell-diameter is unrelated to organ size. Epidermal cell-length is not quite so indifferent, and a negative coefficient appears in the last line of Table 6. Negative coefficients were otherwise conspicuously rare in this enquiry. That the epidermis should respond by division to conditions which other tissues meet by enlargement of their cells is consistent with observations by many workers (e.g. Houghtaling, 1940) that mitosis commonly goes on longer in the epidermis than in other layers. Although the systematic study of cell-sizes becomes much more difficult when the vascular tissues are concerned, Sinnott was able to make an interesting comparison between the vessels and the phloem fibres of the petiole. It was found that phloem fibre diameter was rather strongly correlated both with the general size of the petiole and with the dimensions of other types of cell; in particular there was a very high correlation (0·812) between phloem fibre diameter and pith cell diameter. Vessel diameter seemed to show a more independent behaviour; the comparison of correlations was unfortunately not carried to completion, but the question arose whether the size of the vessels might not be related to the physiological needs of the lamina rather than to local conditions in the petiole. As a measure of physiological need Sinnott chose the fraction: (volume of lamina)/(cross-sectional area of petiole). This showed a significant positive correlation (+0·459) with vessel diameter but a negligible correlation (−0·18) with phloem fibre diameter. This tends to bear out the suggestion that the vessels, unlike the other cells of the petiole, are materially influenced in their behaviour by conditions external to the petiole itself. One consequence of this is that physical proximity of the tissues is not a safe guide to similarities in their behaviour: the diameter of the phloem fibres in a section can be more accurately predicted from measurements of pith cells (correlation 0·812) than from measurements of vessels (correlation 0·575) although the latter are so much nearer to the phloem, and so much more closely related to it histologically.

A treatment which is based on the measurement of mature cells tends to assume a static quality which is uncongenial to most modern morphologists, and observation of specimens at various stages of development is now the usual procedure. Even without much refinement in the mathematical analysis, developmental observations can throw a good deal of light on the true nature of various structural distinctions. We may begin with some work on differences of genetical origin.

Some important measurements on the transverse expansion of young stems were made by Houghtaling (1940), who worked with *Lycopersicum*. She was able to examine not only contrasting stocks, each the product of several generations of inbreeding, and showing striking differences in characteristic thickness of stem, but also certain hybrids between them, and the F_2 families raised from those hybrids. Because the diameter of a stem increases continually, it would be meaningless to specify diameter without setting a standard of age. Houghtaling successfully overcame this difficulty by plotting stem diameter (on a logarithmic grid) against distance back from the stem apex. The relationship between parents and hybrid is then approximately as shown in Fig. 37 (see p. 60), where the parent stems yield the continuous curves and the hybrid stem the broken one. Such graphs immediately suggest that a geometric mean will afford a suitable basis for predicting the condition of the hybrid; in fact, if at some specified distance from the apex (of the order of 3 to 20 cm) the characteristic stem diameters of the parents are D and d, the diameter of the hybrid is $k\sqrt{Dd}$ where $0{\cdot}91 < k < 0{\cdot}97$. This implies that the arithmetic mean of the parent diameters must always significantly exceed the hybrid diameter.

In order to apportion the responsibility for stem expansion between the different tissues the areas occupied by pith, vascular cylinder, and cortex, were separately determined in a series of sections, and graphs were plotted (again logarithmically) showing the variation of each tissue against total area of section. It emerged very strikingly that the logarithmic plot of pith area against stem area has a constant gradient of unity for all the genotypes examined. The percentage of the stem section which is occupied by pith depends upon the genetically-determined characteristic diameter of stem. It is 30% in the thinnest-stemmed variety but 49% in the one with the stoutest stem. The percentage does not vary, however, during the expansion of an individual stem within the

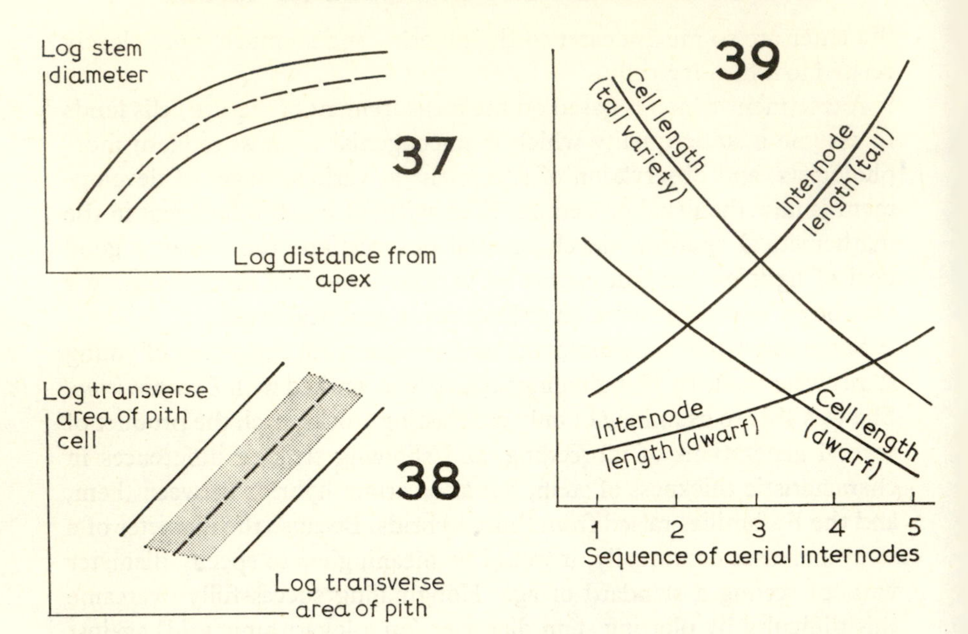

FIGURES 37–39. *Graphical relationships arising in studies of cell-size (simplified from the originals and not to scale).*

37. Relationship in tomato varieties between diameter of young stem and distance of measured stem section from apex. Continuous lines represent two genetically distinct races, the broken line their hybrid (Houghtaling, 1940). **38.** Allometric treatment of the pith in Houghtaling's tomato stems. Continuous lines represent two contrasting inbred parents, the broken line their F_1 hybrid, and the shaded area the range of potential sites for the graphs of F_2 progeny. **39.** Comparison of cell and internode lengths in plants of dwarf and tall wheat varieties (Nilson, Johnson & Gardner, 1957). In so far as greater internode length arises from progression up the stem, it is associated with shorter cells, but in so far as it arises from the genetical properties of the tall variety it is associated with longer cells.

range of the observations. Such a relationship must eventually be overthrown in any stem where a large bulk of secondary vascular tissue is formed, but Houghtaling's material was too young for the consequences of cambial activity to be clearly apparent. The observation that a genetically thicker stem will have a larger fraction of its cross-section occupied by pith is in agreement with the general experience of plant anatomists in many families of dicotyledons, but it is satisfactory to have a demonstration that a hybrid will in this respect fall between its parents. The fact that the proportion of the stem occupied by pith remains constant during an extended period of early life is interesting, but we must not too hastily conclude that it implies a specific physiological coordination. The vascular cylinder constitutes during the same period of development an increasing fraction of the transverse section, and the cortex a diminishing one. It is therefore possible that the constancy of relative pith-size is more or less coincidental; the vascular system is gaining in relative bulk, as its permanently meristematic character would lead one to expect, but as the growth rate of the cortex is declining in about the same measure the line of the epidermis is not for some time affected. In this respect Houghtaling perhaps did not sufficiently allow for the way in which the system of arithmetic might constrain certain relationships to appear. To plot other variables against cross-sectional area of the stem was in any case an expedient of questionable value, because, as between varieties differing greatly in specific thickness, it leads to comparisons between stems which are not at corresponding stages of development.

More importance can be attached to some direct measurements of cell-sizes in transverse section. In Fig. 38 the two continuous lines are graphs of pith-cell area against pith area in developing stems of two varieties. In the logarithmic plot these lines have unit gradient, which means that each variety has a specific number of cells in its pith. The growth shown here is pure cell-expansion. The discontinuous line is derived from the hybrid, which has a specific number of pith cells intermediate between the parental numbers. The shaded area centred upon the hybrid line represents a crowd of graphs, not represented individually, obtained from the F_2 plants. A similar study of cortical cells produced an essentially similar set of relationships. Observations on the epidermis, however, revealed a continuance of cell-division at the same time. More detailed examination of F_2 plants showed that there were genuine differences between them in specific cell-numbers, and that there were positive correlations between cortical cell-number, pith cell-number, and stem diameter. Within the range of these breeding experiments, differences in size appeared to be due primarily to differences in cell-number. Other comparisons, however, showed that this is not invariably the case, and that some extreme thick-stemmed variants are distinguished by the greater size of their cells.

Another study on the genetical control of cell-size was made by Bindloss (1942) who examined tall and dwarf varieties of ten species of cultivated dicotyledons. Most of these, after preliminary investigation, were dismissed as unsuitable for developmental study of cell-size in relation to height. The desirable qualities for this purpose were considered to be the possession of a dominant main axis which would permit accurate measurement of overall height, rapid growth, and a large embryo which could easily be removed from the seed, the intention being to examine development from the earliest stage of germination. Dwarf and tall varieties of tomato and *Zinnia* were chosen for detailed investigation. A refinement not universally practised in such researches, but undoubtedly of considerable value, was the standardization of seed weight. Seeds were weighed individually, and only those in a specified modal range of weight were used. The difference in weight between the tall and dwarf seeds selected in this way was negligible in tomato and very small in *Zinnia*; it is most unlikely that the observed differences between tall and dwarf plants can owe anything to differences in initial size of the embryo.

Bindloss adopted an unusual method for recording the sizes of cells. Whereas most workers have determined cell-sizes by a sampling procedure, and established the position of each estimate within the plant by reference to some independent measurement such as the distance of the sample area from the apex of the shoot, Bindloss chose to measure every cell in an unbroken series from the surface of the apical meristem downwards into the young pith, and she listed a measurement not as having been taken a certain distance back from the apex but as the axial length of the 47th cell in line, and so on. This represents a substantial improvement in technique; if the cells of two varieties differ in size it is obviously possible to argue that cells at equal distances from the apex are not morphologically comparable, because a given distance involves a greater number of mitoses in the smaller-celled variety. Cells the same distance from the apex will thus be at different points along the sequence of events from apical initiation to final differentiation. The Bindloss method appears to offer a much more reliable standard. The 47th cell in a giant may be taken as a fair equivalent for the 47th cell in a dwarf, and it is of much less significance that they are situated, owing to different rates of expansion, at different distances from the apex.

Complete sequences of cells were recorded in tall and dwarf plants at various ages from the dormant embryo onwards. The sequences in the older plants extended to more than 100 cells, and the results were remarkably consistent in the two species and also through the sequence from apex downwards. In tomato and *Zinnia* alike, in the tissues of a very young seedling (seventh to ninth day of germination) each cell of a tall variety is longer than the corresponding cell of a dwarf; in both species, however, this difference is very soon reversed. By about the twenty-first day of germination the dwarf is very clearly characterized by the possession of larger cells, and this state of affairs was maintained as long as observations continued. The position in the dormant embryo was less clear-cut; in tomato the genetically dwarf embryo had consistently larger cells than the tall one, whereas in *Zinnia* there was little difference either way. As there is in any case a general increase in cell-length away from the apex, it is not surprising that the genetic differences became more apparent in the lower part of the sequence, and could not always be recognized in the first few cells of the row. It was clear from inspection of apical sections that dwarf and tall varieties do not markedly differ in the size of the apical meristem. The control of cell-size is strictly related to the individual cell and does not arise incidentally from a change in the total bulk of tissue. It became evident also that even when cultural conditions are such as to make genetically 'tall' and 'dwarf' plants grow to equal heights, the characteristic differences of cell-size are perfectly maintained.

It is not yet possible to offer a final interpretation of these observations, but it seems clear that we should concentrate our attention mainly upon what appears to be the equilibrium condition, in which the dwarf form possesses larger cells than the tall ones. This could rationally be attributed to a difference in mitotic rate, for if the dwarf cell has more time between successive mitoses it will be able to grow larger. The temporary reversal in the younger seedling then presents a difficulty, but it is inherently probable that at that stage a larger part of the total growth which is observed should be due to cell-expansion. The enlargement of existing cells is an especially prominent feature of early seedling life, and such expansion at a time when the tissues of the tall plant had not yet attained their characteristically higher rate of mitosis would go some way towards providing the required explanation.

That the relationship between cell-size and cell-number may be affected in different ways, even by agencies which produce similar effects upon the size of the plant, is sufficiently shown by the work of Nilson, Johnson & Gardner (1957). The subject here was the flowering shoot or 'culm' of wheat, which consists usually of six internodes, the lowest being subterranean. In the mature culm the five aerial internodes increase in length progressively towards the apex (Fig. 39). Over the same range, the lengths of individual cells show a decreasing trend; both epidermal cells and cells of the internal parenchyma were measured, with concordant results. It might appear, therefore, that a situation existed comparable with that reported by Bindloss, in which the longer organ had shorter cells. This relationship only holds good in wheat, however, when the differences in internode length are occasioned by differences of position in the culm. In so far as Nilson and his collaborators followed Bindloss's practice by comparing internodes which differed in length for genetical reasons they found, as shown in Fig. 39, that dwarf varieties were distinguished by shorter cells, which is the reverse of Bindloss's experience with tomato and *Zinnia*.

Although successive leaves or internodes along a shoot are presumably as likely to differ in the balance between cell-division and cell-expansion as in any other respect, there have been relatively few observations on such variation. Technical standards which are adequate for the presentation of a general account of development will not necessarily meet this more exacting demand, but a convincing example can be seen in the work of Sunderland (1960) on the plumular shoot of *Helianthus*. The comparison here was between the tenth leaf of the seedling, which is part of the spiral sequence of leaves characterizing the adult plant, and average figures for leaves three and four, which are almost opposite. Cell counts were made by a maceration technique. There is a considerable period of almost logarithmic increase in cell-number, after which the mitotic rate, hitherto roughly constant, falls rather suddenly to a low level. It is not possible to fix exactly the position of the turn in the cell-number curve, but from Sunderland's graphs it seems that we shall not be far out in saying that the mitotic rate falls in a lower leaf when the area of that leaf is about 55% of its final value, but that the corresponding figure for the tenth leaf is about 77%. In the original text the true magnitude of this distinction is partly obscured by the stress which is laid upon the point of absolute cessation of mitosis in the lamina. This point, which one would think must in any case be very difficult to locate with precision, comes, according to Sunderland, very late in the history of the sunflower leaf. He says that after the point of cessation there is very little further expansion. This may well be so, but it is less interesting in the present connection than the period before the point of cessation (amounting to about twenty days) in which the mitotic rate has been very low. During this period the lower leaf increases its area by about 45%, the upper one by no more than half that amount.

Higher standards of numerical accuracy can be attained by the use of plastochron concepts, as illustrated by the work of Maksymowych (1963), who used the plastochron scale developed for the leaf of *Xanthium* in a study of the growth of cells in the palisade layer and upper epidermis. In the embryonic condition (plastochron age of leaf from −2·5 to zero) both types of cell have almost the same dimensions, the mean area of a cell in a paradermal section being then about 50 μm^2, its height in the thickness of the leaf about 10 μm. At about the zero-point of the plastochron scale the two tissues begin to display divergent modes of growth, the palisade cells increasing mainly in height, the epidermal cells mainly in area. All the essential relationships are shown in Table 7

TABLE 7

Relative growth-rates per plastochron $\frac{1}{x}\frac{dx}{dP}$ for cells of *Xanthium* leaf at intervals of 0·5 plastochrons (extracted from Maksymowych, 1963)

Plastochron age of leaf	Dimensions measured			
	Palisade cell height	Epidermal cell height	Palisade cell area	Epidermal cell area
0·0	0·19	0·01	−0·05	0·15
0·5	0·25	0·02	−0·24	0·42
1·0	0·29	0·05	−0·28	0·69
1·5	0·31	0·09	−0·06	0·96
2·0	0·33	0·10	+0·06	1·18
2·5	0·36	0·12	+0·47	1·15
3·0	0·36	0·13	+0·74	0·76
3·5	0·27	0·11	+0·64	0·39
4·0	0·14	0·10	+0·42	0·15
4·5	0·05	0·07	+0·24	0·06
5·0	0·02	0·04	+0·11	0·03

which gives relative growth rates (per plastochron) at half-plastochron intervals through the main period of cell-expansion. The numbers here have been extracted from a much more comprehensive presentation in the original. In assessing them it must be appreciated that relative growth rates possess a degree of comparability which is independent of the absolute magnitude of the increasing dimension. A value of unity in any column would imply a rate of increase equivalent to doubling during a plastochron interval. This rate is reached for a short period in the area growth of epidermal cells, but we must not be too hasty in seeing here the highest growth rate in the whole table, because area rates are not directly comparable with the linear rates given in other columns. The numerical values here are too large to be treated as infinitesimals (compare, for example, the coefficients of thermal expansion of a metal), but the calculation remains a very simple one; to find the linear growth rate equivalent to an area rate of 1·18 we subtract 1 from the square root of 2·18, and so on. It appears therefore that the linear growth rates in the plane of the leaf lamina are greater, but not enormously greater, than those in the thickness of the leaf. (It must be remembered that these are cell growth rates; tissue growth rates are many times greater in the plane of the lamina, but most of this difference is attributable to cell-division.)

Table 7 shows that height growth is greater in palisade than epidermis down to plastochron age 4, and that area growth is greater in epidermis than palisade down to plastochron age 3. The column for palisade cell area, however, displays certain peculiar features: there is an early period with negative growth rates, which is intelligible only as an indication of rapid anticlinal division, and a later period in which active area growth of the palisade cells is the most conspicuous growth activity remaining in either tissue. In short, it appears that if we think of the growth of the palisade in terms of two phases, a phase of cell-division followed by a phase of cell-expansion, then both of these phases must be delayed relatively to the corresponding phases in the epidermis. Observations by other workers have generally tended to the same conclusion. Thus Haber & Foard (1963) made a study of tobacco leaf, in which there is a good allometric line relationship between leaf width and leaf length; at various points along the allometric line they expressed the number of cells in epidermis and palisade as percentages of the numbers to be attained in the mature state. Table 8, compiled from their graph, suffi-

TABLE 8

Growth of tobacco leaf: length and breadth of leaf in millimetres, numbers of epidermal and palisade cells as percentages of final numbers (observations of Haber & Foard, 1963)

Leaf length	Leaf breadth	No. of epidermal cells	No. of palisade cells
12	4	20	2
22	10	30	10
32	16	40	20
65	30	100	60
85	45	100	100
150	100	100	100

ciently shows the different timing of epidermal and palisade development. Whereas Haber & Foard determined cell-numbers from intact portions of leaf, Maksymowych proceeded by macerating samples of lamina and counting the isolated cells in a haemocytometer. This has the disadvantage that only a total cell count is practicable; in the embryonic stages all distinction between tissues is lost. It was possible to show, however, that the total number of cells in the *Xanthium* leaf, after a period of compound interest growth, comes rather suddenly to a final value. Although theory would lead one to prefer a curved fit, it is good enough in practice to plot cell-number logarithmically and use two straight lines. From a stage at least as early as plastochron age -3, cells divide at a steady rate, about one mitosis every two days; this process continues to age $+3$, and then stops, cell-number thereafter remaining constant. There is thus a rather close coincidence in time between the virtual cessation of mitotic activity and the peak values of palisade cell growth rates shown in Table 7. This accords with the general belief that active cell-expansion is incompatible with the continuation of a meristematic function. The epidermal cells are of course too few in number for their earlier expansion peak to make any perceptible impression upon the 'all-tissues-average' mitotic rate.

The authors whose work has so far been considered in this chapter, while recognizing the facts of apical growth and the histological distinction of tissues, have not been primarily concerned with problems arising from local concentrations of growth activity. The existence of such concentrations, however, must be expected to have the same importance in the study of cell-sizes as it has in other aspects of morphology. Provided that the differences in growth are not very small, much can be discovered by methods which hardly rise above the descriptive level. For example, Whaley & Whaley (1942) were able to identify in the leaf of *Tropaeolum* the points of development at which two particular genes exercise their characteristic influence upon the shape of the mature lamina. The doubly recessive plant has a lobed foliar primordium, maintains an almost uniform distribution of laminar growth, and consequently exhibits a lobed leaf at maturity. The two dominant alleles operate by inducing additional mitoses very early in development, and specifically on those radii of the lamina which constitute the sinuses of the lobed primordium. When the additional cells expand, the result is actually an excess of growth on the sinus radii; if this additional growth were efficiently coordinated in the radial direction the doubly dominant leaf would have 'reversed' lobing, with prominences *between* its principal lateral veins. In reality the excess growth provoked by the dominant alleles is very imperfectly polarized, and its tangential component throws the leaf into undulations. The lamina will no longer fit against a flat surface: if the lateral veins be pressed against such a surface the excess tissue on what were originally the sinus radii will form convex folds, and the shadow of the leaf upon the test plate will be smoothly orbicular.

More conspicuous and more persistent localizations of growth occur in those stems where meristematic activity is so sharply restricted to a particular portion of an internode as to constitute an intercalary meristem. An extreme example is the aerial flowering shoot of *Eleocharis*, described by Evans (1965). The inflorescence shaft here consists of a single distinctive internode, containing large intercellular ventilating chambers of a kind not found in the rhizome system below. This internode often reaches a length of 40 cm, yet at no time does any significant amount of cell-division take place more than 7 mm from its base. The various tissues are distinguished not so much by their maximum rates of mitosis as by the alacrity with which they abandon mitotic activity on moving away from the internode base. The first tissue to stop dividing is the parenchyma destined to form the widely-spaced transverse diaphragms of the ventilating system. These cells cease to divide when only 1·2 mm from the internode base. The tissue which is most persistently mitotic is the photosynthetic parenchyma of the cortex, in which the last mitotic figures are about 7·2 mm from the internode base. Evans has the interesting observation that any selected vertical file of cells in the meristem will generally show a conspicuous periodicity. Moving along the row from a group of short cells resulting from recent divisions we observe a gradual increase in size, then two or three cells in various mitotic stages, then another group of product cells, and so on in regular rotation. This presumably means that a mitotic wave is travelling along the row. Neighbouring files of cells, however, are out of phase, so that there is not necessarily any resultant periodicity of the meristem as a whole.

In *Avena*, which other grasses are likely to resemble, it has been shown by Kaufman, Cassel & Adams (1965) that the intercalary meristem is not exactly at the base of the internode but a little way above it. Above the meristem the epiderm cells as originally produced undergo the unequal 'secondary' mitosis which yields the alternation of long and short cells so characteristic of the mature epidermis of the grasses. In the part below the meristem this unequal division does not occur, short cells are not formed, and the original cells merely elongate to an extraordinary extent. This means, among other things, that this basal portion of the internode has no stomata, which in grasses can arise only from the short type of epidermal cell. Despite these complications, the general behaviour of the meristem is not very different from that of *Eleocharis*. Again there seems to be no general periodicity in the mitotic rate, though some of the photographs are suggestive of the more local type of periodicity suggested by Evans. Again division in the parenchymatous tissues appears to be more active, or perhaps merely more prolonged, in the layers near the surface than in those near the centre. The American authors show a disposition, in their discussion, to generalize the principle that mitosis ceases early in central tissues; they may well be right, and the idea is of course not restricted to intercalary types of meristem.

The one form of localization of growth which is universally characteristic of shoot systems is the specialized activity at the apex of the stem. Here the balance between cell-division and cell-enlargement appears as a property of a stream of differentiating cells which flows out from the rear end of the apical meristem. Although this cell-flow is a real movement, very few workers have recorded any attempt to view it directly. Most of our knowledge is based upon inference from observations which were taken from specimens in an essentially static condition. The much simpler conditions which prevail in roots, however, have facilitated a more dynamic approach, and important theoretical developments have arisen from this experience. Some consideration of these ideas is a necessary part of our discussion. The mathematical relationships which arise are in no way particularly associated with roots, but must exist wherever a localized meristem projects a stream of cells outwards across its own boundary, and the application of the dynamic method to stems must naturally follow as soon as the technical difficulties can be surmounted.

All the significant advances can be seen in the work of Goodwin & Stepka (1945) on the root tip of *Phleum*. Considering only the epidermis, the cells of which are in longitudinal rows, and concentrating our attention upon a single plane of transverse section at a distance x from the tip of the apical meristem, the cells in any row will be streaming across that section in the direction of increasing x. It is convenient to adopt the convention that a cell is 'at' the distance x from the apex at the instant when its basal transverse wall passes across the section. Let r_x be the velocity of a transverse wall at its passage across the section. The distinctive character of the work springs from the fact that Goodwin & Stepka were able to determine this velocity directly. Under the microscope they could see the movement of the wall and obtain an estimate of r_x in microns per mm in no more than ten minutes or so, which is for all practical purposes as good as an instantaneous determination. From their observations they obtained a graph of velocity (Fig. 40, upper part), and hence, by drawing tangents, the graph of dr_x/dx against x. This differential, except that it is expressed in clock time rather than plastochrons, is the exact counterpart of the relative elemental growth rate already discussed in connection with the work of Maksymowych (p. 42). The transverse cell-wall serves the same purpose as Maksymowych's ink-mark, and the growth rates obtained are essentially tissue growth rates. We see that the forward thrust of the root is largely generated in a surprisingly thin transverse slice centred about 375 μm from the apex, but the observations so far introduced can give no information on mitotic rates or cell growth rates. Let L be the mean length of epidermal cells at the transverse plane x. The graph of L against x (Fig. 40) is sigmoid with a very gentle rise, the graph of dL/dx consequently very broad-topped, with a poorly defined maximum at about $x = 650$ μm.

The root must be presumed to be in a state of dynamic equilibrium; averaging over a period, the number of cells in that part of a cell-row which lies between the apex and the plane x is to be maintained constant. Therefore there must be one mitosis in this file of cells during the time which one cell takes to pass completely through the plane x. Goodwin & Stepka undoubtedly possessed the resources to measure this transit time directly, but they perceived that such an observation was unnecessary, and preferred to calculate the interval from graphs already prepared. At the moment when the basal wall of an average cell is passing through the plane x with velocity r_x, the apical end is in the plane $(x - L)$ and has consequently a velocity $r_{(x-L)}$. We require the average velocity of the apical cell-end during its passage from $(x - L)$ to x. Strictly this ought to be obtained by integration, but because L is small in relation to x the acceleration can without serious loss of accuracy be treated as uniform, giving an average speed of $\frac{1}{2}\{r_x + r_{(x-L)}\}$. Having the distance and the average speed we know the time of the journey, which will be equal to the interval between two successive mitoses in the cell-row. The rate of division will be the reciprocal of the time; between the apex and the plane x there are $\frac{1}{2L}\left\{r_x + r_{(x-L)}\right\}$ new cells formed per minute. Just as with other quantities, if we require an elemental rate of cell-production to show how the production of new cells is shared between different transverse slices or elements of the root, the general rate just obtained must be differentiated with respect to x. As a serviceable approximation to a considerably more complicated expression, Goodwin & Stepka arrived at:

$$\frac{1}{L}\left\{\frac{\mathrm{d}r_x}{\mathrm{d}x} - \frac{r_x}{L}\frac{\mathrm{d}L}{\mathrm{d}x}\right\}$$

as the elemental rate of cell-production for the epidermis, and this is graphed in Fig. 40. It is not a mitotic rate in the ordinary sense of that expression, being related to length of cell-row rather than to the number of pre-existing nuclei. The lower part of Fig. 40 nevertheless represents in principle a partitioning of tissue growth rate into a mitotic component and a cell growth rate component for every region of the epidermis. Such an analysis represents an impressive approach to finality. It does not deal with questions of statistical variability, but so far as the description of average values is concerned nothing further remains to be discovered. Goodwin & Stepka extended their ideas in a splendid way to the inner tissues of the root, and were able to calculate elemental rates for the production of new cell-wall area. The whole investigation clearly foreshadows the likely course of development in research techniques for other organs.

Another study of root growth based on velocity measurements was undertaken by Hejnowicz (1959), who worked with wheat and took his observations by photographing the epidermal cell-pattern every forty minutes. In his analysis he introduced the idea of 'linear cell density', the number of cells per mm along the length of a row. This is the reciprocal of cell-length, and has for each tissue a maximum at some characteristic short distance from the tip. These considerations bring out the interesting point that the mitotic rate (which can be expressed in terms of number of new cells produced per existing cell per hour) is substantially constant through an appreciable length of root, but that the 'frequency of cell-division' (new cells produced per mm length of cell-row per hour) is then directly proportional to linear cell density. This is an effective warning against reliance on subjective impressions: the mitotic rate is not necessarily greater in those parts of a tissue where mitotic figures appear to be most abundant.

In principle there cannot be any reason why the parts played in the

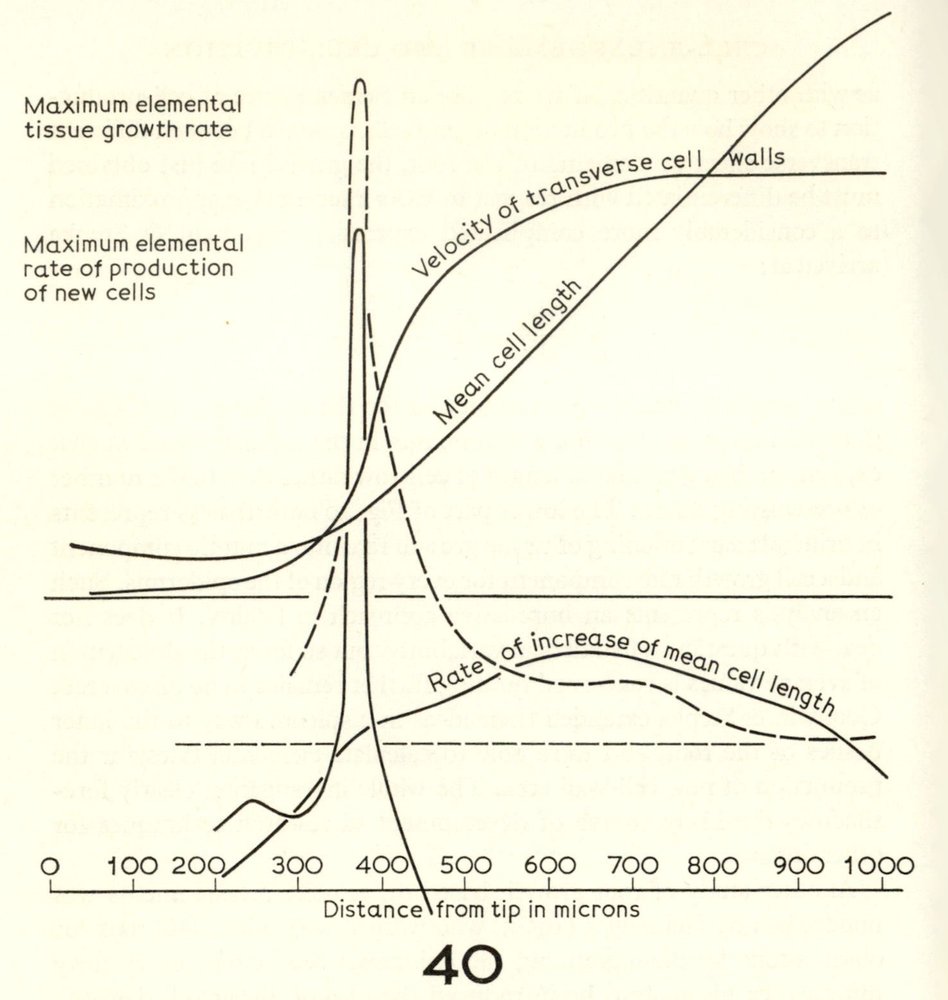

FIGURE 40. *The growth of an organ investigated by measurement of the velocities produced. Plotted from data of Goodwin & Stepka* (1945).

The base-line (applied at three different levels for reasons of clarity) shows distance from the tip. All the curves are waveforms which travel through the tissue in the wake of the advancing apex. Those most directly related to observation are plotted to the uppermost base-line and give the distributions of velocity and of mean cell-length. Differentiation from the velocity distribution yields the graph of elemental tissue growth rate (shown in broken line on a separate base). The two graphs on the lowest base-line show how responsibility for this tissue growth is to be partitioned between cell-enlargement and the production of new cells. Calculation of the elemental rate of cell-production by the method explained in the text (p. 71) reveals a single very short wave of mitotic concentration which accounts for the main peak of tissue growth. Simple differentiation from the observed distribution of cell-length generates a long flat-topped wave of cell-enlargement which alone sustains the residual growth of tissue in the region 450–800 μm from the tip.

growth of an organ by cell-division and cell-enlargement should not be ascertained, as in the preceding examples, by direct observation and measurement of cells. Except for the manipulative difficulties which may be presented by a particular class of specimen there is no agency by which these matters can be hidden from us, and no need accordingly for any development of technique beyond the level of straightforward visual recording. Manipulative difficulties can however be formidable, and in particular they usually prohibit sustained observation upon a single specimen of any internal tissue. Conclusions have to be based on a comparison of specimens arrested in various phases of growth, and the analytical treatment of the observations must then attempt a separation of general principles of growth on the one hand from non-developmental individual peculiarities on the other. The perilous and uncertain quality of such an analysis gives a strong inducement to exploit any method by which the tissues can be caused to retain some permanent trace of their developmental history.

It has long been evident that the production of a sectorial chimaera affords an opportunity to determine the part played by cell-division in the growth of an organ. Such a chimaera will arise whenever a cell undergoes a change which is transmissible by mitosis. If, for instance, a cell becomes polyploid while its neighbours remain diploid, and if further mitoses take place subsequent to the change, a patch (or sector) of polyploid tissue will be formed, and the size and shape of that sector will give a valuable clue to the processes of growth in the area concerned. It is inconvenient to use any type of change which can be detected only by cytological examination, and fortunately genetic markers can be found which render the chimaeral sectors visible without special treatment. The work of Stein & Steffensen (1959) depended upon the availability of *Zea* seeds heterozygous for a chlorophyll deficiency, the gene concerned being near the end of a chromosome arm. It was known from previous work that a terminal deficiency of this chromosome arm is satisfactorily transmitted through the mitotic cycle. The technique therefore was to soak the heterozygous embryos in water, to expose them to a standardized dose of X-rays, and to study the length and distribution of the yellow streaks in the mature leaves after the seedlings had established themselves. In this form of experiment the chlorophyll-deficient cells are those from which the dominant allele has been omitted because of breakage in the appropriate chromosome arm. It is desirable in all such work that the radiation dosage should be kept as low as is consistent with the production of a reasonably adequate number of chimaeral sectors. An excessive dosage will too often result in the simultaneous modification of two or more cells within a small portion of tissue, giving rise to a sector which is derived from several parent cells but will be falsely attributed to one, so leading to an incorrect estimate of the mitotic rate. Furthermore the object will usually be to determine the conditions of growth in the normal organ, and the argument is likely to rest upon the assumption that the modified cells do not greatly differ from the unmodified ones in their rate of division. Such an assumption may be justified where the selected genetic marker is the only genetic difference which has to be considered, but a less favourable situation would arise if any significant proportion of the chimaeral sectors were to be suffering from other, uncontrolled, forms of chromosome damage. Homozygosity of the material for genes other than the marker pair will be a considerable safeguard against undesirable complications; even so it will be better if the occurrence of cells with multiple chromosome breakages can be kept to a minimum. Marker genes often require defined conditions to produce their full effect. Stein & Steffensen were obliged to move their plants into a cool room at a certain age to forestall the production of chlorophyll in the chimaeral sectors. In a matter of this kind it is easy for a single research team to achieve satisfactory standardization of its own procedure, but difficult to ensure that results will be correct in any absolute sense because any failure to recognize all the sectors is likely to operate differentially against the smaller ones.

In an irradiated embryo the leaves of the plumule are at different stages of development. As Stein & Steffensen were merely measuring the lengths of the pale stripes in their leaves with an ordinary ruler they disregarded sectors shorter than 1 mm, and thus required a considerable number of mitoses to have taken place between treatment and observation. Leaves 1 and 2 had sectors too short for them to measure, and they confined their attention to leaves 3 to 6 inclusive. The use of a grass leaf, in which the cells are arranged in regular rows, makes it possible to simplify the problem by concentrating on growth in length and disregarding growth in width. In *Zea* there appears in fact to be a very uniform distribution of transverse growth across most of the surface; except for an occasional broader stripe near the margin the observed sectors differed little in width. It also seems clear that the genetical change involved in the production of a chimaeral sector was not associated with any gross disturbance of growth; mutant and non-mutant tissues showed no obvious difference in cell-size, nor was there any evidence of shearing stresses between them such as must arise where adjoining tissues grow at different speeds.

Even in these favourable conditions, however, the analysis of experimental results is not free from difficulty. If the radiation treatment is correctly applied the distribution of the original chlorophyll-deficient cells will be random, so that the number of sectors observed in a leaf at maturity ought to be proportional to the number of cells capable of further division which were present in the primordium of that leaf at the time of irradiation. Stein & Steffensen found, as this argument would suggest, that the number of sectors was greatest in the lowest leaf, which would be the largest primordium when irradiated, and diminished progressively in subsequent leaves. They did not however claim, and were clearly not in a position to demonstrate, any simple mathematical relationship. The larger primordium has a smaller number of mitoses remaining to be performed, which means that the sectors in the lower leaf will be shorter than those of the upper leaf. In this respect it is disturbing to find that no less than 45% of the recorded sectors in leaf 3 lay in an apical region where the average sector length seems to have been no more than 2 or 3 mm. Remembering that any sector shorter than 1 mm would pass unrecorded, and making a reasonable allowance for statistical variation of sector-length (an important point upon which the original text gives no information), there are clearly grounds for suspicion that a significant proportion of sectors may have been omitted from the count. The effect of this would be to reduce the variation in sector-count between successive leaves, so that it would under-represent the original differences of primordium size. Taking the published figures, the sector-count of leaf 3 exceeds that of leaf 6 by a factor of only 5·5, though the primordia differed in length by a factor of 28; in tissue which was all essentially embryonic, one may well question the reality of the fivefold difference in cell-length which this calculation would demand.

The length of a sector indicates the number of mitoses occurring between irradiation and maturity. A region with many short sectors is one which had many cells in the primordium and a low rate of future mitotic activity; a region with fewer but longer sectors consisted originally of fewer cells but had a more extended course of development before it. The observations of Stein & Steffensen are amply sufficient to show that in the leaf of *Zea* there is a marked difference between the apical part of the leaf, where sectors are numerous but short, and the basal part, where sectors are longer but less frequent. As a first approximation, the apical part of the leaf ceases meristematic activity at an early age, and is carried upwards by the basal part in which cell-division continues much longer. Abandonment of mitotic activity afterwards spreads progressively downwards from the already passive tip into the lower regions until eventually the growth of even the lowest part of the blade is brought to a standstill. On referring the observations back to the situation when the X-rays were applied, it appears that the distinction between the base and tip is already assuming some importance in primordia not much more than 0·1 mm in height. It is an inherent limitation of the method that such an estimate must be based upon comparison between two successive leaves. When the primordial leaf 6 is irradiated at a height of only 0·05 mm, the sectors in the basal part of that leaf at maturity are longer than those in the apical part by a factor of about 7. We cannot deduce from this that the primordium as irradiated had any significant distinction between base and apex at all; the measurements merely show that the apex must have been distinguished by a lower mitotic rate during some period subsequent to treatment. We can fix this period more precisely only with the aid of observations on lower leaves. Leaf 5, which was about 0·15 mm high when affected by the radiation, yielded apical sectors considerably shorter than those in leaf 6, whereas the basal sectors were similar in both leaves. A cell near the apex of a primordium 0·05 mm high has a sufficient period of meristematic activity still before it to produce a strip of tissue 2·5 cm in length. By the time the primordium reaches a height of 0·15 mm the final cessation of mitosis has drawn appreciably nearer, and the potential output of the cell has fallen to less than 1 cm of tissue.

The method of Stein & Steffensen can thus provide useful information at a descriptive or roughly quantitative level. There are several difficulties, however, in the way of any attempt to reach higher standards of accuracy. Because the sectors at the time of observation commonly reach a length exceeding a quarter of the length of the leaf itself, it is necessary to enquire what point in the length of the sector is to be regarded as its point of origin. The original presentation of results was based on an arbitrary decision that the position of a sector should be taken as the position of its basal end. This is obviously impossible to justify. In principle one would seek to refer back the meristematic potential indicated by the length of a sector to the position in the primordium at which the parent cell of that sector was standing when the

X-rays hit it. The form of calculation required for this would be a development of the concept of elemental growth rate and would evidently call for a very complete series of observations. As individual plants may differ in the size of their adult leaves it is necessary to combine upon a single base-to-apex scale observations from leaves with differing rates of growth. The American investigators assumed that differences in the leaf length would mostly arise from differences in the growth of the basal portion, and that it would be sufficient to combine the data according to the distance from the leaf-tip. This perhaps gives a standard of accuracy consistent with that attained in other aspects of their work, but it is only an approximate solution. There are objections also to a procedure which examines a series of leaves along a plumule rather than successive stages at a selected node. The most serious limitation, however, arises from the length of time which passes, and the great number of mitoses which must occur, between treatment and observation; the pattern of growth cannot be understood in detail while its effects are recorded only as long-term summations. The use of genetic markers would assume much greater importance if a shortening of the experimental period could be coupled with efficient detection of the necessarily smaller chimaeral sectors produced.

Perhaps the most informative use of genetic markers is that in which a patch of mutant tissue is established in the apex of a stem. Treating a seedling in this way Steffensen (1968) was able to show that the apical dome of maize goes through a series of changes; a mutant cell low in the dome will work its way out and affect only the lower leaves. Also it appears that the dome may rotate upon the stem, the mutant sectors forming a slow spiral in the distichous sequence of leaves. Such observations, though interesting, tend to emphasize the primarily qualitative status of the method.

Attempts to determine the rate of cell-division have sometimes been based upon experimental interference with the mitotic cycle. Thus Denne (1966) exposed young leaflets of *Trifolium* to colchicine for an eight-hour period, the object being to accumulate metaphase figures in the various tissues. Any cell which comes into the metaphase condition during the colchicine exposure will be prevented from proceeding further, and upon subsequent examination the mitotic rate of every tissue can be calculated from the proportion of its cells which have entered metaphase during the hours of colchicine application. The principle is unexceptionable, and there is no doubt that such techniques can yield estimates of mitotic rate which will be adequate for many purposes. Whether the method can achieve the standard of numerical accuracy which Denne apparently attributed to it is a different question altogether, and the value of the *Trifolium* investigation is materially diminished by the failure to maintain any effective control of the errors involved. Basically the situation was closely comparable with that studied by Maksymowych in *Xanthium*. The layers of the primordial leaf consisted of cells all of similar dimensions. Absolute cell-number per layer increased logarithmically for a considerable period (Fig. 41, see p. 79) and then rather suddenly became constant; the palisade tissue continued its course of logarithmic increase rather longer than the other tissues. Mean cell-area in the paradermal plane remained approximately constant during the logarithmic phase, but thereafter increased rapidly. All this is clear enough, though the graphs, instead of being scaled in plastochrons, are reckoned in days from the 'initiation' of the leaf, which is a rather indefinite zero-point, and the growth rates which are put forward are obtained by some method, not fully disclosed, which does not seem to have been as reliable as Maksymowych's careful numerical differentiation. The direct observations on number and size of cells are not however linked by calculation to the colchicine work. Denne's presentation therefore gives the reader no adequate assurance that the colchicine-based estimates of mitotic rate are compatible with the increase in cell-number determined by other means. In the absence of that assurance, and having regard to the uncertainties concerning such matters as the rate of penetration of colchicine into the tissues, it is impossible to feel complete confidence in the fairly small differences of mitotic rate by which the tissues are said to be distinguished.

The experimental interventions just described are not of a very radical nature, the applied treatments being essentially only tools which aid in the observation of normal growth. More fundamental changes in the character of the work arise when treated material is permitted to continue along a disturbed course of development, with the object of gaining insight as to the underlying physiological relationships. A strikingly successful experiment of this kind was that of Haber (1962), who raised plants from wheat seeds which had been given a dose of Co^{60} gamma-radiation so heavy as to produce almost total mitotic inhibition. The plants from irradiated seed grew slowly for ten or twelve days, by which time they had reached the end of their capabilities; without mitosis, nothing further could be done. A treated plant at its limit is similar in

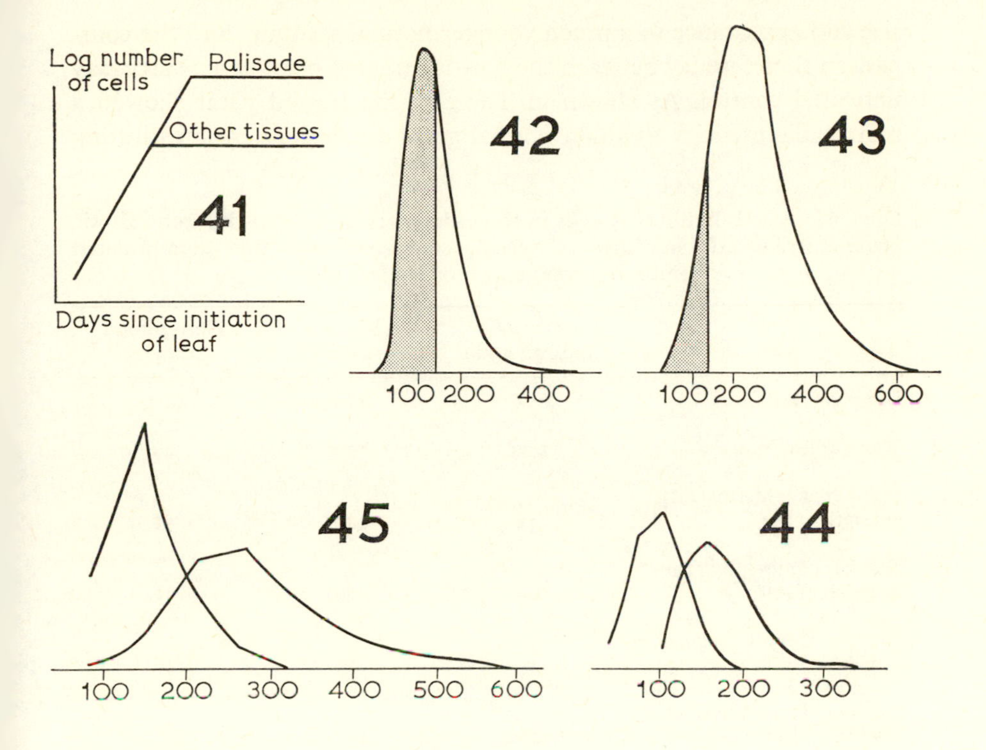

FIGURES 41–45.

41. Scheme claimed by Denne (1966) to operate in leaflet of *Trifolium*: compound interest of cell-numbers up to a sudden cessation which comes later in palisade than in other tissues. **42–45.** Frequency distributions relating to analysis of the data of Brotherton & Bartlett (1918) as explained in the text, pp. 80–86. In every case the base-line shows size-classes of primary cell outlines, the height of the curve the percentage of cells falling into each class. Class-interval is 30 μm for normal material but 60 μm for etiolated specimen, so that graphs **43** and **45** enclose twice the areas of **42** and **44**, a purely technical point without influence on the conclusions to be drawn. **42 & 43.** Total primary cell populations of normal and etiolated epicotyls respectively. Shaded areas denote cells less than 140 μm long and are respectively 59% and $10\frac{1}{2}$% of the total, as against the original claim of 59% and 15%. As a method of estimating the length which qualifies a cell for division this is open to objection because it falsely identifies all cells below the critical length with cells which have not divided. **44 & 45.** Corresponding graphs in which divided and undivided primary outlines have been plotted separately. The vertical line of section is here replaced by a triangular area of overlap which represents the zone of probability within which some divided primaries may be shorter than some undivided ones. The peak of the overlap triangle (at 135 μm for the normal plant and 200 μm for the etiolated one) marks the only length which has zero diagnostic value as between divided and undivided primaries, being an equally probable length for both classes.

size and appearance to a much younger normal seedling, and the comparison therefore lay between the ten-day treated plant and a three-day untreated control. As shown in Table 9, the treated plant showed a remarkable capacity to attain normal gross dimensions by substituting

TABLE 9

Dimensions and numbers of cells in three-day normal and ten-day non-mitotic (gamma-irradiated) seedlings of wheat, compared with the ungerminated embryo (observations of Haber, 1962)

	Embryo in unsown grain	Normal seedling	Non-mitotic seedling
Length of leaf (mm)	0·66	18·2	18·1
Width of leaf (mm)	1·23	2·8	2·6
Number of epidermal cells in length of leaf	52	149	51
Number of mesophyll cells in length of leaf	49	429	51
Number of epidermal cells in width of leaf	137	133	146
Number of mesophyll cells in width of leaf	128	138	120

a quite abnormal degree of cell-expansion for the lost alternative of cell-division. The leaf evidently has an ability to control its own shape by a mechanism which is independent of the very existence of individual cells, or of the localized growth phenomena involved in the production of stomata and epidermal hairs, these developments being excluded by the exposure to cobalt.

In a significant proportion of cases the development of a tissue involves two (sometimes perhaps more than two) separate periods of active mitosis: 'primary' cells are formed, undergo a more or less prolonged adult life during which they may enlarge very greatly and differentiate in various ways, and are then launched again into further mitotic sequences. Considerations of wall thickness alone are sufficient to ensure that this compound type of developmental history will often leave visible traces in the organs at maturity. The relationships are necessarily more complicated than those which arise where there is only a single peak of mitotic activity; we may take as an example the epicotyl of *Phaseolus multiflorus*. Here the epidermal cells as first formed have pointed ends, and although many of them at a later stage undergo transverse division into two or more 'secondary' cells the pointed 'primary' outlines can still be distinguished in the mature stem. An attempt to exploit the opportunity so presented was made by Brotherton & Bartlett (1918). Their procedure was defective, the primitive statistical ideas available to them affording no sufficient basis either for the proper planning of the observational work or for an efficient analysis of the results obtained. With a little recalculation, however, we may extract from their records some valuable information about the problems arising in these complex tissues.

There had been some dispute as to the growth of etiolated plants; it was not clear whether an etiolated internode was longer merely because of increased cell-expansion, or whether there was also an increase in the number of cells. Brotherton & Bartlett began by taking samples of normal and etiolated epicotyls, and determining the frequency distribution of length for each class. Etiolated epicotyls averaged 305 mm and normal ones 85 mm, an increase by a factor of about 3·6. They then selected for anatomical examination the longest normal epicotyl available (141 mm), but an etiolated specimen which was only 372 mm in length; this is much less than 141 $\times$ 3·6, and considerably shorter also than the longest observed etiolated specimen at 517 mm. Their argument was that an effect of light in retarding cell-division would be conclusively proved 'if an etiolated epicotyl of less extreme position in the variation curve should contain more cells'. It is now easy to see that their decision to stake everything on a comparison between two chosen specimens was fundamentally unsound. Instead of increasing, as they supposed, the security of their conclusions regarding the effects of etiolation, it merely ensured that any difference in cell-number, no matter from what cause, would be attributed to etiolation. As they seem to have taken no special care to secure genetical homogeneity in their material, and did not even sow seeds of standardized weight, the work cannot be taken seriously as a physiological experiment. It did, however, incidentally involve a very thorough survey of the distribution of epidermal cell-lengths, both primary and secondary, through the two chosen epicotyls.

The American investigators clearly appreciated that conditions must be expected to vary along the length of the internode, and accordingly divided each epicotyl into ten equal segments, keeping separate records of cell-size for each segment. In the event, the regularity of the measurements was hardly sufficient to justify such a fine subdivision; without sacrificing anything of importance we may simplify our own discussion by recognizing three regions only: the basal three-tenths of the epicotyl, a middle portion comprising the fourth to seventh tenths inclusive, and the apical three-tenths. At no point in the original text is any disposition shown to question the exact morphological equivalence of corresponding subdivisions of length in internodes grown under different regimes. Yet this matter of equivalence must surely be one of the main sources of difficulty. If one stem has had its length increased by the operation of some agency which does not affect all parts of the internode equally, but which stimulates growth differentially near one end, then the tissue at mid-length of the internode so modified is not, in a structural sense, the counterpart of that at mid-length of an unchanged internode. The figures given in the first three lines of Table 10 show how morphological

TABLE 10

Mean lengths in microns of primary and secondary epidermal cells of the epicotyl of *Phaseolus* for normal and etiolated plants, and the increase factors associated with etiolation (recalculated from observations by Brotherton & Bartlett, 1918). 'Primary cells' here includes those primary outlines within which secondary division has taken place

	Region of epicotyl		
	Basal	Middle	Upper
Mean length of primary cells: (normal)	91	153	151
(etiolated)	237	258	266
Increase factors	2·60	1·69	1·76
Mean length of secondary cells: (normal)	—	87	83
(etiolated)	106	121	154
Increase factors	—	1·39	1·85

equivalence becomes an issue in the Brotherton & Bartlett investigation. The basal section of a normal epicotyl is distinguished by a failure of its primary cells to attain the length of those at a higher level; in the etiolated epicotyl (and it is really quite irrelevant to our present interest whether its condition was purely an etiolation effect) the primary cell-lengths are more nearly uniform. The increase factors in the third line of the table merely restate the position; whatever increased the length of the etiolated stem has had a greater effect on the basal section. If these changes in size occurred without any change in cell-number it would follow that there had been a flow of cells out of the basal section. Any cell which would, in the normal stem, have stood 29% of the way up the epicotyl, would be thrust beyond the 30% mark by the disproportionate expansion of those below it, and would figure, in the etiolated stem, as a cell of the middle section. So long as no question of a difference in the distribution of cell-divisions arises, equations can readily be set up to express the magnitude of cell-flow resulting from any given distortion in the distribution of cell-sizes. In the present state of knowledge a mathematical exercise of this type would be quite sterile botanically; further consideration of Table 10 will illustrate the difficulties which stand in the way of any complete solution. The two selected epicotyls differed in length by a factor of about 2·64. The primary cells of the basal segment showed an increase factor of 2·60, which is very nearly sufficient. This coincidence escaped the attention of Brotherton & Bartlett but it might very naturally lead to the suggestion that the base of the epicotyl was producing the extra length by increased primary cell-extension, extra divisions at the primary stage being found only in the middle and upper sections where cell-extension falls short of the required ratio. In reality there is no evidence at all to support such a conclusion; the observations of cell-length would be perfectly consistent with the contrary hypothesis: that the necessary increase in total primary cell population was entirely due to enhanced meristematic activity in the internode base, accompanied by upward flow of the products. Furthermore the uncertainties of this situation are unrelated to the magnitude of the increase in primary cell-number, and do not disappear as that increase approximates to zero. Two internodes with equal numbers of cells might have strongly contrasting distributions of mitotic activity in the primary meristem. It therefore seems necessary to reject both the assumption that morphologically equivalent points can be found by corresponding fractions of internode length and the assumption (clearly implicit in the original) that measurement of cell-size could be pursued to a point which would render counts of cell-number unnecessary.

The lower part of a normal epicotyl is hardly concerned at all in the second period of cell-division, that in which secondary cells are formed. The rarity of secondary cells in this part of the stem accounts for the blanks in Table 10. Conditions governing this secondary division are evidently very different from those affecting the primary meristematic phase. In the normal epicotyl the lengths of secondary cells are rather uniform throughout the upper parts; in the etiolated specimen, however, we see not only the abundant occurrence of secondary divisions in the lower part, a qualitatively new phenomenon, but also a marked gradient, with considerably longer secondary cells towards the apex. It is impossible to derive an increase factor directly for the basal region, but if we were to attribute to this part of the normal stem the secondary cell-length which prevails higher up we should obtain a notional value of 1·25. Whether we use this device or not, the last line of Table 10 shows that the main impact of etiolation (or of whatever factors were inadvertently associated with it) falls, so far as the elongation of secondary cells is concerned, upon the upper part of the epicotyl. This, being directly opposite to the corresponding result for primary cells, calls for further consideration.

It is strange that Brotherton & Bartlett, after laborious practical work, and after taking such care to separate the various parts of the epicotyl, never carried their analysis to the point we have now reached. The disappointingly static quality of their discussion is manifestly a consequence of this failure. Because they did not effectively compare the upper and lower parts, they could visualize neither the time-sequence nor the spatial distribution of the events responsible for the configurations they had measured. The distribution of cell-lengths in the normal epicotyl in fact points strongly to a partial segregation of primary and secondary types of division into different regions of the stem. Suppose primary division, initially of general occurrence, to be arrested in the upper levels while continuing at the base. The lower primaries, if they exceed a certain length, divide into new primaries. An upper primary, however, will pass through a period during which elongation cannot be obscured in this way by primary division. Hence the greater length of upper primaries. Suppose further that, after primary division has everywhere ceased, the opportunity to divide in the secondary manner is granted to every cell which has attained a certain qualifying standard of length. The lower primaries, having been until very recently engaged in primary division, will have had little time for this additional elongation, and will generally fail to satisfy the condition for secondary division. Upper primaries, on the other hand, will have had an interval between the cessation of one mode of division and the commencement of the other, and may have enlarged significantly during this respite. Upon such a theoretical scheme is it possible to impose any single modification which would account for the peculiarities observed in the etiolated epicotyl? So far as the figures in Table 10 are concerned, it would be sufficient merely to postulate a greater rate of cell-expansion at every stage. Rapid expansion in the primary phase would enable some cells to qualify for addition primary divisions, so increasing the total number of primaries, as required. Rapid expansion in the interval between the two phases would not only make the primary outlines longer, which is one of the effects to be produced, but could also enable the lower primaries to qualify for secondary division, perhaps the most important point of all. The characteristic mode of cessation of primary division, progressively from the top downwards, would be unchanged, and would account for the observed gradients in both primary and secondary cell-lengths. This outline is too simple to be completely plausible, yet it seems to offer a reasonable basis for further investigation.

The only stage of development which Brotherton & Bartlett attempted to consider in isolation was the initiation of secondary division. They introduced a 'specific mean length for division, which is simply the mean of the lengths at which division takes place in a large number of cells', and claimed to have shown that this specific mean length was about 140 μm in normal and etiolated specimens alike. They seem not to have realized that this statement, if true, would have remarkable implications for the expansion of a secondary cell after its initial formation. It would mean that a secondary cell in the normal epicotyl would be expanding from 70 μm to about 85 μm (which is an increase of only 21%) but that a secondary cell in the etiolated stem would have to expand from the same 70 μm standard to about 126 μm, which is an increase of 80%. That etiolation or any other agency should have a fourfold effect on growth after the secondary division but no detectable effect on growth before it does not seem particularly likely, and suggests the advisability of re-examining the grounds for the assertion of the invariable specific length.

The original calculation was based upon frequency distributions, one for the normal and one for the etiolated stem, each derived from the lengths of a thousand primary cells (divided and undivided without distinction). In the normal stem sample 59% of these primaries were undivided, but in the etiolated sample only 15%. Brotherton & Bartlett sought to evaluate their 'specific mean length for division' by making appropriate cut-offs from these curves as shown in Figs. 42 & 43. They said that the shortest 59% of normal primaries had lengths below 140 μm; that is, the shaded area in Fig. 42 is 59% of the whole area under the curve. Now this procedure involves the quite indefensible assumption that the 59% of primaries shorter than 140 μm and the 59% which have not undergone secondary division will be the same 59%. It does not allow for the possible consequences if there should be undivided cells longer than 140 μm or divided ones shorter than 140 μm. Also, while a 59% cut-off is near enough to the centre of the distribution to be accomplished with some pretence to accuracy, the 15% cut-off required in connection with the etiolated specimen is a much more hazardous operation. Fig. 43, in which the original data yield a cut-off at 140 μm of only 10½% instead of the required 15%, not only reveals a mistake in the 1918 calculation but shows the steepness of the curve, and consequently the vulnerability to observational error, in the region concerned.

All this procedure was in any case perfectly unnecessary. Let us define the specific length for division, in terms more acceptable to modern statistical taste, as the length which has the same probability for a divided as for an undivided cell. If now we plot, in Figs. 44 & 45, the frequency distributions for lengths of divided and undivided primary cells in normal and etiolated specimens, the required specific lengths can be read off directly from the points of intersection as approximately 135 μm in the normal stem and 200 μm in the etiolated one. The secondary cells are thereby called upon to expand by about 26% in both cases.

Evidently the attempt to understand the growth of a tissue as complex as the *Phaseolus* epidermis demands not only an adequate supply of observations but also a willingness by the investigator to calculate separately every single item in the train of events. The work of Brotherton & Bartlett shows a certain lack of vision in this respect. To take but one further example, although they had the data at their disposal, they appear never to have calculated the quotient: (length of divided primary cell)/(length of secondary cell). The values are about 2·07 for the normal plant, but 2·21 for the etiolated one. This ought to mean that a secondary division additional to the first one is three times as common in the primary units of the etiolated stem. An analysis which does not descend to this level of detail must remain largely ineffective.

Chapter Four

The Succession of Parts

In the case which may be considered most typical of the vascular plants a shoot produces leaves and branches in a sequence which seems potentially endless. In many species we know of nothing which could prevent a single stem apex from being so maintained (of course in carefully controlled conditions) as to keep up a steady output of new lateral organs in perpetuity. It is logical therefore to begin with the consideration of a somewhat idealized uniform flow of new parts.

At the tip of a shoot, in the arrangement of young leaves around the apex of the stem, we encounter geometrical configurations which attracted attention long before there was any possibility of placing their study upon a sound scientific basis. The early literature of phyllotaxy contains much accurate observation, but the theoretical developments, in so far as they progressed beyond a formal classification of the geometrical patterns discovered, retained until quite recently an ineffective and semi-mystical quality.

The modern treatment essentially begins with the consideration that in a transverse section through the apical region every young leaf is moving outwards, approximately along a radius of the system, and is at the same time increasing in size. The simplest assumption (indeed the only simple assumption available) will be that the shape of the primordium changes only slowly. Our scheme of geometry will therefore be based, as a first approximation, upon a principle of similarity; angles will remain unchanged while linear dimensions increase. To find out what kind of framework is needed to satisfy this requirement it is necessary to resort to the use of polar coordinates as shown in Fig. 46 (see p. 88). The position of a point P is here to be specified by $r =$ OP, the radial distance of P from the centre O, and θ, the angle which its radius makes with a fixed radius for which $\theta = 0$. The radius OP is swinging in the direction of the arrow and elongating at the same time. The point at the end of this moving radius describes a curved trajectory ST, the curvature being such as to maintain a constant value for the angle between radius

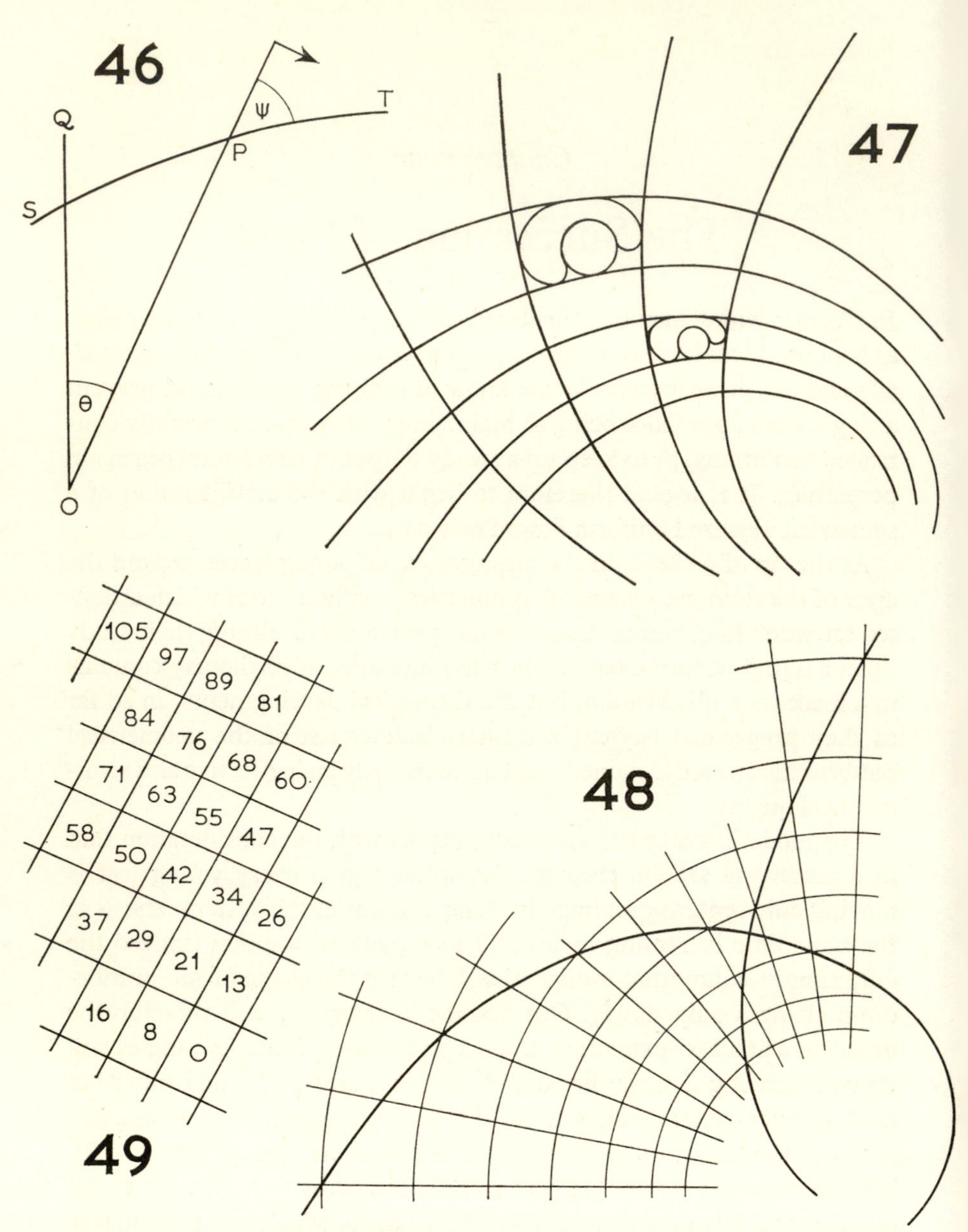

FIGURES 46–49. *The geometry of spiral phyllotaxy.*

46. Definition of a logarithmic spiral. OQ being fixed OP revolves about O in the direction shown by the arrow, P moving along the curve ST, which is to be made a logarithmic spiral. If *r* is the distance of P from O, *r* and θ are both increasing but ψ is constant. Movement of P towards T resolves into two com-

and curve. Expressing this fixed-angle condition as a differential equation we have:

$$\frac{1}{r}\frac{dr}{d\theta} = \cot\psi = \text{constant.}$$

Upon integration this yields:

$$r = ae^{b\theta}$$

where $b = \cot\psi$ and a is a new constant. This is the polar equation of a logarithmic or equiangular spiral. In order to reproduce the essential character of a phyllotactic system we must construct a network of such spirals, as in Fig. 47. Here a set of equally spaced logarithmic spirals of uniform inclination is crossed by another set winding in the opposite direction but with a different angle of slope. Because of the equiangular property, all the meshes of such a design are of the same shape, however much they may differ in size. By this method of purely geometrical construction many of the appearances which are characteristic of shoot systems can be duplicated. Once the spirals have been drawn in the correct numbers, and with the appropriate angles of slope, it remains only to sketch the outlines of the primordia in the meshes; ideally we might do this by photographing a single drawing, and printing the negative to a different scale of enlargement for each space in the chart.

---

ponents: dr along OP away from O, and $rd\theta$ perpendicular to OP away from Q. **47.** Principle of geometrical similarity. Two sets of logarithmic spirals wind in opposite directions about a common centre, the spirals in each set being identical and uniformly spaced. Then the meshes (and such biological forms as may be inscribed within them) differ in size but not in shape. **48.** Graphical construction. The groundwork consists of a set of equally spaced radii and a set of concentric circles, the circles being progressively closer as the radii converge, so as to give a net of similar cells. Logarithmic spirals are obtained by combining rotation with radial movement in any agreed proportions; of the two shown one advances one space towards the centre for two steps of rotation, the other three for two. These shapes were taken off for construction of the previous diagram. **49.** Rule for serial numbering of parts. Number 0 is arbitrarily chosen, other numbers assigned by the rule: in moving one step up any row, add the number of rows in the set of rows of which the chosen row is a member. Sets of 8 and 13 have been used in construction, but other sets can be followed with varying degrees of difficulty. The reader should satisfy himself, for instance, that 16–21–26 continues, and will be found making another turn across the face of the specimen as 71–76–81, and is one of a set of five, another member of that set being 37–42–47 . . .–97, and so on.

Many authors have embellished their discussions of phyllotaxy with diagrams prepared in this way. The accurate construction of a logarithmic spiral is however a very tedious process calling for repeated reference to tables, and an approximate method has often been used, with or without acknowledgement. In Fig. 48 the straight lines are equally spaced radii. The concentric circles are not equally spaced but are drawn further apart as the radii diverge, so as to make the meshes between radii and circles as nearly as possible constant in shape. In practice this may be done by eye, without actual measurement. Now a circle and its radius represent the two limiting cases between which every logarithmic spiral must exist. A circle has no radial growth but represents a pure rotation, while a radius has no rotation but only radial extension. We require a curve which will combine rotation and growth in fixed proportions, so we must move across the circle/radius grid by a fixed rule: so many meshes outwards, so many tangentially. The spirals in Fig. 48 have been drawn by rules respectively of (3 and 2) and (1 and 2), and closely approximate to the equiangular form. It is generally convenient to transfer the shapes generated by this method into other diagrams by the use of templates.

Logarithmic spirals give the framework appropriate for the development of leaves on the apical surface of a stem, which we have so far treated as flat. Where the spirals run down on to the sides of an essentially cylindrical stem, they become deformed into the shape of a helix or screw-thread. Intermediate states, in which the spirals traverse a conical or domed surface, necessarily also exist. Regardless of these differences, the existence of a regular frame permits us to attach serial numbers to the leaves. The rule will be, that in moving upwards along a spiral the serial number is to be increased by the number of spirals in the set of which the selected spiral is a member. Fig. 49 shows the commencement of the numbering operation in a system which has eight spirals in one direction and thirteen in the other; one leaf having been arbitrarily designated as 0, the rest follows, subtraction being needed to locate some of the lower members. The numbers assigned in this way indicate the order of development of the parts. Should a stem be so constituted as to produce two leaves at a time, then it will automatically ensue that each number will appear twice, which we may deal with by labelling leaves as 3 and 3(*a*), etc., but where leaves are developed only one at a time no duplication of numbers can arise.

Having established the system of serial numbers, we next set up a definition to the effect that any sequence of leaves between which the serial number increases by constant steps is to be known as a parastichy. Inspection of Fig. 49 will show that not all parastichies are equally conspicuous. 16–29–42–55–68 makes a powerful visual impact whereas 8–42–76 would have to be traced with some care. Both are alike in being formed as logarithmic spirals, but one is given special emphasis by the way in which its members meet one another on broad surfaces of contact. Parastichies which stand out in this manner are distinguished as the contact parastichies; a spiral system commonly has two sets of these and we may adopt the convenient notation of Church (1904) and describe the geometry of Fig. 49 as an (8 + 13) phyllotactic system. One other category of parastichy which calls for special recognition is that with a constant difference of unity: 0–1–2–3, etc. This is traditionally known as the genetic spiral, and in systems with numerous closely packed organs it is extremely inconspicuous.

The study of phyllotaxy has from the outset been dominated by the systematic recurrence of certain particular numbers as the constant differences of contact parastichies in widely assorted botanical material. In reproductive and vegetative parts, and in all the major groups of vascular plants, except perhaps the Articulatae, a single mathematical principle appears to be in operation. It has long been recognized that the arrangement of leaves is intimately linked with the peculiarities of certain number-series, all of which conform to the rule that each term shall be the sum of the two preceding terms. Mathematicians have been aware since the thirteenth century that series of this type have some distinctive properties, Leonardo of Pisa having studied the series: 1, 2, 3, 5, 8. . . . This has ever since been known as the Fibonacci series, because (surnames not then being in use) Leonardo was distinguished by a patronymic; 'Fi-' is the equivalent of 'son of'. This series, and a selection of others related to it, are given in Table 11. We can at once set up the following generalizations:

(*a*) A shoot apex which produces one leaf at a time will ordinarily have two sets of contact parastichies, the constant differences of which will be consecutive terms of one and the same series; likely contact parastichy formulations will be (3 + 5), (5 + 8), (4 + 7), etc.

(*b*) Although a particular species may show a preference for uncommon numbers, the Fibonacci series in general predominates, those more remote from it becoming progressively rarer. In a large miscellaneous

TABLE II
Number series encountered in the examination of spiral phyllotaxy

Fibonacci series	1	2	3	5	8	13	21	34	...
Other series starting from 1	1	3	4	7	11	18	29	47	...
	1	4	5	9	14	23	37	60	...
	1	5	6	11	17	28	45	73	...
Series starting from 2	2	5	7	12	19	31	50	81	...

collection (3 + 5) and (5 + 8) will be much more abundant than (4 + 7) or (7 + 11), but some stem succulents are noteworthy because they produce with considerable frequency numbers like 9 which are elsewhere decidedly uncommon, and which belong to the lower lines of Table 11.

(*c*) The numbers become larger as the individual leaf becomes smaller and less important in relation to the whole system. Other things being equal, (2 + 3) would suggest a leisurely development of few large leaves, (21 + 34) a rapid succession of small ones.

(*d*) When the contact parastichy numbers have a common factor, the system contains more than one leaf at each stage of development. (8 + 14), being equal to 2 (4 + 7), implies development of leaves in pairs, and is distinguished as a bijugate system. Bijugate systems are not excessively rare, being normal for capitula of *Dipsacus*, etc., but the unijugate condition is more widespread, and the still higher levels of complexity (trijugate, etc.) are quite outside the experience of most botanists.

There have been two distinct and largely independent historical processes by which the numbers set out in Table 11 have entered the consciousness of botanists concerned with phyllotaxy, and most of the early work falls accordingly into two contrasting styles. One school of thought has regarded contact relationships as a secondary and almost accidental class of effects, and has concentrated on the study of the genetic spiral. From this point of view the first objective in the study of any specimen must be to determine the angular divergence, the angle of rotation round the genetic spiral from one unit to the next. As it is generally impracticable to locate the centre of the system and apply a protractor, it has become traditional to look for an orthostichy, an exact vertical superposition of leaves. An orthostichy would appear in transverse section as a radius, in side view as a vertical line, running through the midpoints of leaves at regular intervals. If two leaves on the same orthostichy are separated by n internodes and if the genetic spiral makes m turns between them, the system has an angular divergence of $m/n \times 360°$. When divergence angles estimated by the orthostichy method are expressed as fractions of a circle most of them fall into fraction series closely related to the number series already introduced. These fractions are given in Table 12, where it will be seen that the Fibonacci series supplies all the numerators.

TABLE 12

Fraction series encountered in the estimation of angular divergences. The parent number-series are those of Table 11

	Observed fractions					Exact final value, as fraction of circle	Approximate angular equivalent
Fibonacci series	$\frac{1}{3}$	$\frac{2}{5}$	$\frac{3}{8}$	$\frac{5}{13}$	...	$\frac{1}{2}(3-\sqrt{5})$	$137\frac{1}{2}°$
Other series starting from 1	$\frac{1}{4}$	$\frac{2}{7}$	$\frac{3}{11}$	$\frac{5}{18}$	...	$\frac{1}{10}(5-\sqrt{5})$	$99\frac{1}{2}°$
	$\frac{1}{5}$	$\frac{2}{9}$	$\frac{3}{14}$	$\frac{5}{23}$	...	$\frac{1}{22}(7-\sqrt{5})$	78°
	$\frac{1}{6}$	$\frac{2}{11}$	$\frac{3}{17}$	$\frac{5}{28}$	...	$\frac{1}{38}(9-\sqrt{5})$	64°
Series starting from 2	$\frac{1}{2}$	$\frac{2}{5}$	$\frac{3}{7}$	$\frac{5}{12}$	...	$\frac{1}{22}(7+\sqrt{5})$	151°

Other writers, including Church (1904), have objected to this whole procedure, saying that no reliance can be placed on angles which cannot be directly measured, and that the identification of an orthostichy is in practice a matter of subjective impression. The force of this objection becomes apparent when the mathematical properties of the fraction series are investigated. It follows from the way in which the numbers are derived that the fractions in any series are successive approximations to a final value which is characteristic of that series. The final values are given at the right of Table 12, where it will be seen that all of them are based on $\sqrt{5}$ and are consequently irrational. Furthermore, as will become clear upon evaluating the approximate fractions as decimals, the convergence towards the final value of any series is quite rapid. The difference between 5/13 and 8/21 of a circle, for example, is so small as to be, in many botanical contexts, quite imperceptible. These relationships make it possible for the objectors to say that if the angular divergence had the irrational final value of any one of the series, so that orthostichies did not exist, an observer who was deliberately seeking an orthostichy in order to estimate divergence angle would merely select a very steep parastichy, thereby extracting one of the approximate fractions of the series. On this view, if one specimen is said to show 5/13 phyllotaxy while another is 8/21, the difference will be attributed, not to a fundamental difference of construction, but rather to variations in the conditions of observation (or the accuracy of the observers). All that has happened is that one man has accepted as a true vertical a parastichy in which the other detected (or thought he could detect) a slight inclination. There seems to be no real prospect of attaining the standard of observational accuracy which would be required to refute this suggestion. A good impression of the difficulties can be obtained from the work of Davies (1939), who took measurements with a graduated circular platform having a central opening through which a twig could be passed. He obtained a mean divergence angle of 137° 39′ 57″, which is near to the final Fibonacci value of 137° 30′ 28″. This was an impressive technical achievement, but a discrepancy of almost a sixth of a degree would be quite enough to invalidate any critical test of phyllotactic theory, and the variability of individual observations was disturbingly large; Davies himself regarded these fluctuations as genuine, but the fact that they were symmetrically distributed about the mean, as most biologically significant phenomena are not, is more suggestive of an error distribution.

Those who have approached the study of phyllotaxy through the counting of contact parastichies have seen the problem in the main as one of packing, of the physical pressure of one primordium upon another. Those who have primarily concerned themselves with divergence angles have tended to adopt schemes in which the initiation of a leaf is governed by hormonal influences radiating from pre-existing leaves. No system of causal explanation has ever commanded general acceptance, and the main effect of such discussions has been to give prominence to the ineradicable defects of both of the traditional forms of presentation. On one side the formulation of divergence fractions very soon reaches a point beyond which actual measurement ceases to be possible. On the other side, Church and other advocates of the parastichy method, although claiming unique qualities of simplicity and certainty for their procedure, are obliged in fact to elaborate their schemes to a degree which many readers find intolerable. Not only is the number of possible contact parastichy formulations very large, but transitions occur, as for example when a tapering fircone passes from (8 + 13) at the bottom to (3 + 5) at the top, with consequent extinction of some parastichy lines. By reasoning which does not seem to be fully recoverable from his text, but which certainly involved questionable ideas about fluid vortices and the slipping of one primordium over another, Church persuaded himself that parastichies of the two contact sets must always intersect orthogonally, that is, at right angles. Orthogonal intersection of parastichies, though apparently a logical necessity to Church, has not appeared so to most other writers, and is in many cases clearly contrary to observation. Church increased the rigidity of his scheme still further by attributing to each contact parastichy pattern its own unique and distinctive divergence angle, and also its own characteristic value of a quantity which he called the bulk ratio. With the aid of a mathematical friend (Hayes & Church, 1904) he was able to elaborate this particular concept in a manner which is enjoyable but not indispensably necessary; stripped of its ornaments the bulk ratio is merely a measure (Fig. 50, see p. 96) of the angle subtended at the centre of the system by each constituent member.

The account offered by Church is the most extreme and unyielding assertion of the fundamental significance of contact parastichies. It is a point of view from which any change in the phyllotactic pattern can be envisaged only as an uncompromising juxtaposition of two incompatible orthogonal frameworks. The resulting irregularities can be catalogued, according to rules of forbidding complexity, but are not to be understood in terms of any gradual change. The application to growing shoots of such inflexible ideas can give no lasting satisfaction.

We owe to Richards (1951) a scheme of phyllotactic description which is capable of accommodating processes of gradual transition and which, for the first time, places contact parastichies and divergence angle in their proper relationship to each other and to the growth of the specimen. Richards shows by means of diagrams similar in principle to Fig. 50 that it is possible to draw very different-looking sets of contact parastichies upon a given genetic spiral, and concludes, as others have done before, that parastichy counts reflect the shape of the parts as much as their arrangement. The significance of this conclusion is more closely

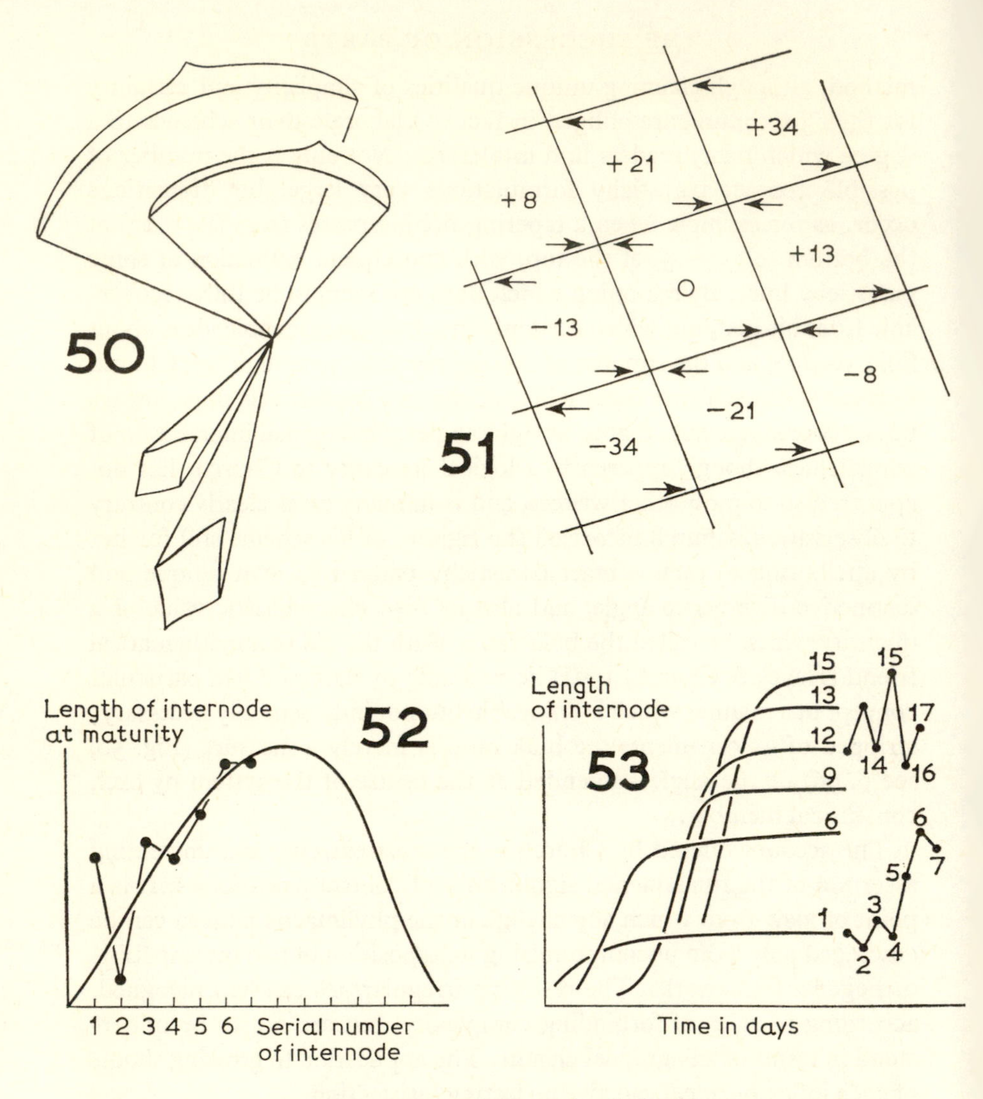

FIGURES 50–53. *Phyllotactic relationships.*

50. Comparison of two spiral systems differing only in bulk ratio, which is high for the laterally expanded organs, low for those of quadrilateral form. The relative positions of the parts in respect both of angular divergence and of distance from the centre, are identical in both halves of the figure, so the genetic spirals are also identical. Differences of contact parastichy formulation must however be expected, in view of the difference regarding overlap of the parts as seen from the centre. **51.** Consequences of change in shape; part of a diagram analogous with Fig. 49, in which each member is to undergo lateral extension as indicated by arrows. Movements will be governed by serial order, e.g. the advancing edge of +13 is to intrude between 0 and +21, not between +21 and +34. Examination will show that extension as indicated will lead to the obliteration of con-

examined in Fig. 51 and the accompanying caption. Richards points out, however, that divergence angle does not, in itself, define even the form of the genetic spiral. To complete the design, some measure of radial expansion is also necessary. He adopts for this purpose the plastochron ratio, which is the ratio (greater than 1) between the radial distances from the centre of the system of any two consecutive members. Richards investigated the trigonometrical relationships between the network of intersecting parastichies and the form of the genetic spiral, defined by the two quantities: plastochron ratio and divergence angle. From this purely mathematical enquiry there emerged a most ingenious doubly logarithmic transformation. A new quantity, the phyllotactic index (P.I.) is set up, such that:

$$\text{P.I.} = 0{\cdot}38 - 2{\cdot}39 \log_{10} \log_{10} (\text{plastochron ratio})$$

The significance of this move is that the resulting phyllotactic index assumes an integral value whenever two sets of parastichies in a Fibonacci system intersect orthogonally. When the (8 + 13) parastichies cross at right angles the index is 5, and so on. A fractional value of the index, such as 5·2, means that no two parastichy sets are orthogonal. Which particular parastichies, in any given situation, are made prominent as the contact parastichies, is a secondary consideration depending on the shape of the parts.

Richards computed tables by which any worker is enabled to surmount the purely mathematical difficulties inherent in a system of spiral phyllotaxy. Given the plastochron ratio it is a simple matter to evaluate

---

tact parastichies with high differences and the establishment of new ones with low differences. For example 0 and −21 will be progressively separated by the intrusion between them of −13 and −8, which will ultimately make contact with each other; when the interface between −13 and −8 has become wide enough an observer will become aware of a set of five contact parastichies. When the process is considered in more detail it will be found that there is a complex interchange of point-contacts and line-contacts, always in the sense unfavourable to the visibility of parastichies with large characteristic differences. **52.** Principle of distribution of internode length (Groom, 1908). Continuous curve shows rise and fall of mature internode length during a season's growth. A subdecussate system of diminishing amplitude has been superimposed upon this. **53.** Subdecussate system in tomato (Thompson & Heimsch, 1964). Growth curves of selected internodes are plotted to show a rising trend equivalent to the upward sweep of the 'seasonal' curve in Fig. 52. By introducing additional observations of mature lengths only, two portions of the subdecussate oscillation have been displayed in full.

the phyllotactic index and then, by reference to the tables, to find the angle of intersection between the parastichies of any specified pair of sets. Conversely, from an observation that specified parastichies cross at a particular measured angle it is equally easy to calculate in the opposite direction to obtain phyllotactic index and plastochron ratio. The treatment offered by Richards is based in the first instance on the transverse plane of a Fibonacci system, but can be extended in two respects to take in other possible configurations. It is sometimes necessary to consider a conical or domed specimen, or for some other reason to take measurements in a plane which is not accurately transverse. Richards meets this need by providing a supplementary formula and table. On any surface which is tilted at an angle to the transverse plane one has to use an 'equivalent phyllotactic index' which is readily interconvertible with the phyllotactic index when the angle of tilt is known. It is also desirable to provide for the interpretation of specimens showing the less common types of phyllotactic spiral. This can be done very simply; the phyllotactic index can still be related to the angles of parastichy intersection. The significance of particular numerical values is changed, so that with parastichy sets such as (4 + 7) orthogonal intersection is no longer associated with integral values of the index, but there is no disturbance of the basic principle that phyllotactic index and the angles of parastichy intersection are mutually interconvertible quantities.

By using this system of calculation it is possible to deal rationally with intermediate states of spiral phyllotaxy. An example can be seen in the study by Rees (1964) of the oil palm (*Elaeis*). Phyllotactic index in transverse sections of the apex of this plant varies from 3·6 to 4·2 with an average of about 3·85, which seems to be almost independent of age and cultural conditions. The equivalent phyllotactic index on the sides of the stem is more variable owing to differences in stem elongation and radial growth, but is still nearer to 4 than to any other whole number over the greater part of the trunk. Values as high as 5·2 can be found at the base of a plant, but must be regarded as transitory. As a general rule, therefore, it is the (5 + 8) parastichies which are nearest to orthogonal intersection. Whether they are also the sets which make the strongest visual impression is another matter; often the (8 + 13), and sometimes the (3 + 5), will be more conspicuous for quite accidental reasons connected with the state of preservation of the leaf bases.

Rees seized upon the connection which must exist, in a system of closely-packed primordia, between the plastochron ratio and the growth rate of the tissues. His average phyllotactic index of 3·85 corresponds to a plastochron ratio of 1·0853. This is the factor by which any length measured in the transverse plane will increase during one whole plastochron. This increase is achieved by continuous compound interest, so that the relative growth rate per plastochron will be $\log_e$ 1·0853, which is 0·0819. In the oil-palm a plastochron is about seventeen days, so the daily rate of growth is 1/17 of the plastochron rate, which is 0·00481 or 0·48% per day. This is a linear rate, and is small enough to be treated as an infinitesimal. Rees asserts with confidence that the longitudinal growth of a primordium cannot in *Elaeis* be more than the transverse growth; owing to the principle of similarity the radial and tangential transverse rates are naturally to be treated as equal. It follows that we may take as an upper bound to the volumetric daily rate in the primordium three times the linear daily rate, say 1·5% per day. If mean cell-size remains constant and all the tissue is meristematic, as may be approximately the case in the early stages, this implies a mitotic cycle of about seventy days, a very slow rate which appears however to be constantly maintained. In this way the work of Richards can be extended to form a complete mathematical pathway from tissue growth rate to the geometry of parastichy systems. For other examples of calculations in which tissue growth rate is related to the plastochron scale see Berg & Cutter (1969).

It is common to find the patterns of spiral phyllotaxy displayed in great perfection, but there are in nature several disturbing influences which can in varying degrees distort and modify the characteristic geometry of a Fibonacci system. In so far as these modifications express themselves in changed angular divergences they must be connected with the questions of symmetry which are to be discussed in the next chapter. Considerations of angular divergence probably ought not, however, to be placed in the forefront of any general study of phyllotactic irregularities. We have to deal essentially with the introduction into a previously steady output of new leaves of an intermittent or spasmodic tendency, and attention must accordingly be directed to the twin issues of timing and of internode length. As soon as it has been demonstrated that the interval between one leaf and the next has become subject to fluctuation, then we have identified a move, however incomplete, towards the production of leaves in whorls; in practice it will be sufficient to consider only whorls of two, and to discuss what may be called a subdecussate system of phyllotaxy.

In a perfect pair of leaves, as seen in a typical decussate system, the members are separated by an internode of zero length, and the plastochron between them is also zero, whether it be measured by calendar and clock or by the correlative method of Fig. 28. It follows that the incomplete pairing of leaves which occurs in less perfectly decussate shoots may be detected by searching for an internode which is shorter than would be reasonably intermediate between the lengths of those above and below. In general the growth of a shoot during a single season is associated with an orderly rise and fall in internode lengths measured at maturity. For a regular spiral or distichous system the graph of mature internode length against serial number will have a smooth form not radically different from a graph of growth rate against time (e.g. middle curve of Fig. 22). In a subdecussate system the long-term rise and fall remain, but a shorter and more localized oscillation is added, as in Fig. 52. This situation seems first to have been systematically studied by Groom (1908, 1909), and is undoubtedly very common indeed. In particular we are obliged to conclude quite generally that any shoot which is genetically conditioned to produce a spiral phyllotaxy, but which starts its growth in some bilaterally symmetrical site (which necessarily includes a leaf-axil or the region between two opposed cotyledons) *must* have impressed upon it some measure of imperfect decussation. No doubt there are many specimens in which the oscillations are of small amplitude and soon die away, the apex then settling down to a steady spiral routine. Subdecussate patterns are however often conspicuous to mere visual inspection, without any need for measurement; axillary branches of *Spartium*, *Sarothamnus*, and *Cytisus* afford convincing examples. More especially in seedlings, the opposite arrangement of the first two leaves is sometimes rigidly standardized, as in *Phaseolus*, where the plumule begins with an opposite pair, though subsequent leaves are all alternate. In seedlings of *Ulmus* Henry (1910) not only found some species to have more persistent plumular decussation than others, but succeeded in obtaining Mendelian ratios for the segregation of this character in hybrids.

Few investigators have taken measurements with the specific intention of studying the subdecussate condition, but subdecussation is so common that its consequences can often be observed in sets of observations taken for other purposes. Reference has already been made (p. 39) to the uneven spacing of the growth curves obtained by Erickson & Michelini; this is a typical piece of subdecussate behaviour. A rather deeper analysis was attempted by Dormer & Bentley (1952); see also Dormer (1965). This more detailed examination of the problem runs almost immediately into mathematical difficulties of considerable severity, and it appears most unlikely that any rapid progress can be made. Two sources of complication can however be identified. In the first place certain leaves, for reasons which are not entirely clear, do not participate in the process of imperfect pairing which is observable among their neighbours (Dormer & Bentley offered a tentative explanation for this in terms of angular divergence: there is an incompatibility between the Fibonacci divergence and a system of right angles). Of much greater importance is the observational discovery that the various developmental processes of each leaf display a large measure of independence in their responses to phyllotactic disturbance. Where two leaves are loosely associated as members of an imperfect pair, some of their physiological mechanisms are moved further towards complete synchronization than others. As a result, while the broad outlines of a subdecussate system can be ascertained without much difficulty, the details will admit of no generalization. Theory can be developed far enough to predict that one plastochron will be shorter than another, but not far enough to predict how much shorter, and the amount of shortening will be different for each quantity we choose to adopt as the basis of our plastochron scale.

The differences which may arise between successive leaves and internodes as a result of phyllotactic disturbance are not necessarily small, and in planning any physiological investigation which calls for quantitative assessment of individual units the phyllotactic state of the material ought to be a matter of concern. To take measurements from single leaves on a plumule with spiral phyllotaxy exposes a research worker immediately to sources of error which may be very large and which nobody knows how to control. Consider for example the measurements of tomato internode lengths given by Thompson & Heimsch (1964) and shown in Fig. 53. Each internode here follows its own course of growth to a final length which is greatest for internode no. 15; to represent so many curves on a common base would be confusing, so only a selection has been given. The shorter graphs at the right, showing mature lengths only, follow a typical oscillatory pattern. Internode 2 is shorter than 1, 4 is shorter than 3, but at that point comes the first of the characteristic phase-breaks of subdecussate systems. Leaf 5 is unpaired and its internode is simply intermediate between those below and those above; the next pair is composed of leaves 6 and 7, and internode 7 is accordingly shorter than 6. This is a very ordinary situation, and the phase-break at leaf 5 could have been predicted beforehand with considerable confidence; nothing can, however, disguise the inherent unsuitability of such a plant for many kinds of experimental enquiry.

Shoots in which there is any reasonably regular pattern of whorled or distichous phyllotaxy appear in general to be remarkably free from the complex disturbances of timing just considered. Distichous sequences of leaves are particularly stable and offer perhaps the most favourable conditions which can be found for quantitative study of the development of single leaves. Verticillate systems are not always quite so straightforward, because whorls are often sufficiently imperfect for there to be a perceptible succession in time among the leaves of a single group; some care may consequently be needed to distinguish strictly equivalent members at successive levels. Even so, uncertainties concerning the timing of developmental processes can usually be reduced to quite small proportions. It does not follow that distichous and verticillate systems are free from difficulty in other respects, but the problems lie mainly in the study of radial symmetry, and for the present we need give detailed consideration only to some specialized examples among the monocotyledons. In these plants peculiar modifications of phyllotaxy can be found which are rare or unknown in other groups, and which merit some attention even at a purely descriptive level. One such system is the spirodistichous pattern (Fig. 54, see p. 103) in which the two ranks of leaves, instead of being straight, wind in slow helices.

The most constructive approach to all these unusual systems is probably through the study of the early stages, attempting to find the origin of the phyllotaxy in the circumstances of the plumule or axillary bud. Whereas in dicotyledons the first leaf of a plumule generally stands at right angles to the plane of the cotyledons, so giving rise in the majority of cases to a spiral or decussate system, in most monocotyledons the first plumular leaf stands at 180° to the single cotyledon, a situation to which a distichous pattern appears to be a natural sequel. A corresponding relationship often exists in the axillary case; the lateral shoot of a dicotyledon usually has its first leaf at 90° to the axillant leaf, but in a monocotyledon the first leaf of the branch is between the branch and the parent axis, of which configuration the lemma of the grass spikelet is perhaps the best-known example. Haccius (1952) described some monocotyledons in which the first leaf of the plumule is not opposed to the

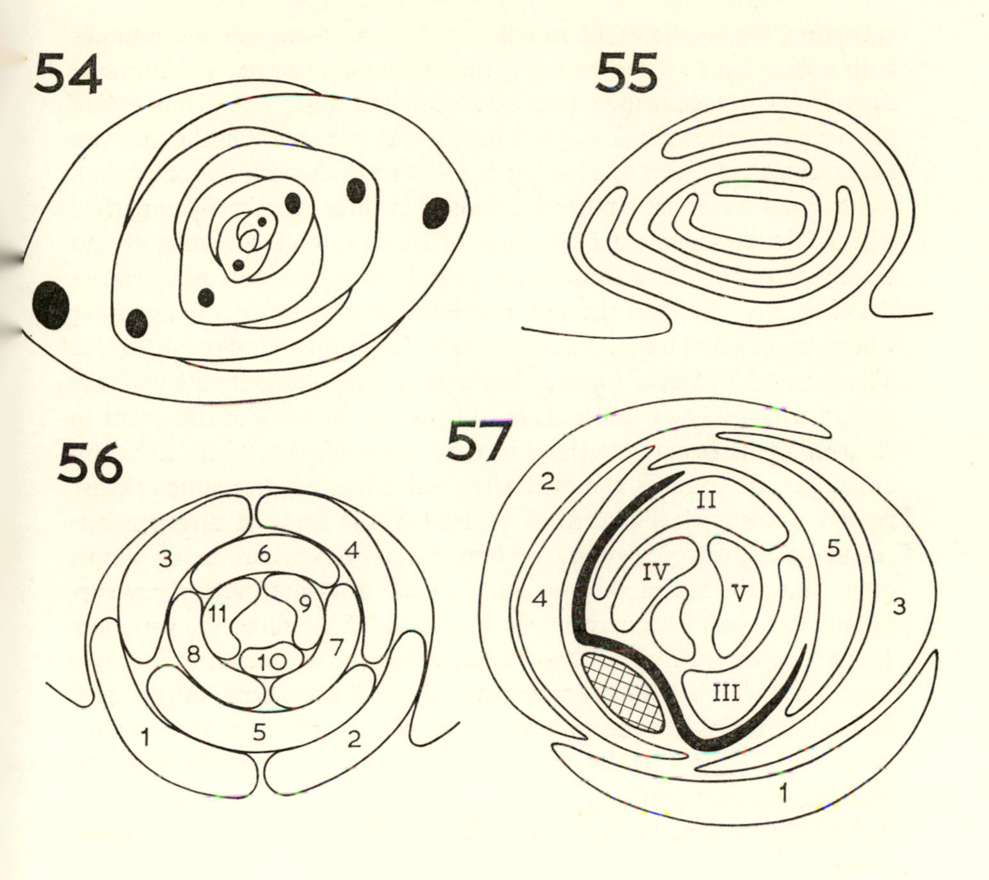

FIGURES 54–57. *Phyllotaxy in monocotyledons.*

54. Spirodistichous system with slowly-winding ranks of leaves. **55.** Seedling of *Enalus*, plumule with a strongly dorsiventral distichous succession of leaves in the basal groove of the cotyledon. **56.** Commonest state of the plumule in *Stratiotes*, again in a groove of the cotyledon. Imperfect whorls of three involving irregularities of angular divergence and with a change in direction of spiral between leaves 2 and 3 (Haccius, 1952). **57.** Continuation of phyllotactic system from parent shoot to branch as observed by Charlton (1968). A spirodistichous pattern of rather rapid twist, leaves of the parent shoot forming the two sequences 1–3–5 and 2–4. The cross-hatched structure is the direct continuation of the parent shoot, its apparent lateral position being deceptive. The other parts shown are leaves of the axillary bud of leaf 5. To some of these Roman numerals have been assigned, but the first one, which is a specialized prophyll, is distinguished only by being shown in black. The parent shoot and the branch together present the spectacle of a single spirodistichous system with two sequences 1–3–5–II–IV and 2–4–(apex + prophyll)–III–V.

cotyledon. Fig. 55 shows the situation in *Enalus*, where growth continues with a distichous system which is markedly dorsiventral. In *Stratiotes* the normal arrangement, found by Haccius in thirty-one plants out of forty-eight, is shown in Fig. 56, where it will be noted that the first two leaves come very near to forming a whorl of three with the cotyledon. Subsequent leaves are similarly disposed in three-membered imperfect whorls, the developmental sequence of leaves in each set being always readily detectable. The state of the mature shoot amounts to spirotristichy, because the lines through corresponding members of successive whorls are of spiral form: one such line is that through leaves 3, 6, 9, . . . The positions of leaves 1 and 2, however, are not conformable with the system which follows. If we were to consider the state of this plant in terms of a genetic spiral we should have to conclude that (*a*) the direction of the genetic spiral is reversed after leaf 2 (since 2 lies anticlockwise from 1 but 4 is clockwise from 3), and (*b*) the angle of divergence is subject to regular oscillation, the first leaf in each 'whorl' being almost opposite to the last of the previous 'whorl'. It is not particularly surprising that the initiation of such a system should be subject to variation. In two of her plants Haccius found a sustained production of two-membered whorls, and she was inclined to interpret some of the other anomalous forms she encountered in terms of a delayed transition from two-membered to three-membered whorls, a transition which, in the 'normal' plumule, takes place almost at the outset. This view finds a certain justification in the fact that axillary shoots of *Stratiotes* commonly produce as many as four pairs of almost decussate scale-leaves before going over to spirotristichy.

In some of these specialized phyllotaxies reversal of the spiral occurs quite regularly. A good example can be seen in the work of Charlton (1968) on the shoots of *Echinodorus*. Vegetative propagation here proceeds by means of a 'pseudostolon', which on superficial examination appears lateral but is in reality a direct continuation of the erect shoot which forms the main leaf rosette. Upon the pseudostolon, the leaves are small scales in 'false' (meaning merely imperfect) whorls of three. The scales of each whorl are fused into a common sheath, and the internodes between successive whorls are long, but the order of succession of the leaves within a whorl is always obvious, and every whorl after the first two shows a reversal of spiral as compared with its predecessor. In such imperfect whorls there is often a relationship between the position of a leaf and its potential for development of an axillary shoot. Thus in the pseudostolon of *Echinodorus*, after some irregularities in the first two whorls, each whorl has only one axillary bud, that in the axil of its first leaf. In *Stratiotes* it is common for axillary shoots to appear only at every sixth leaf, that is one in every alternate whorl of three leaves. On monocotyledonous rhizomes and other specialized propagative shoots a noticeably regular spacing of lateral branches appears to be commonplace. Thus in *Butomus* (Weber, 1950) laterals are often spaced at intervals of eight internodes. Similar effects may be observed in the aerial shoot systems of *Cupressus* and other conifers.

Another aspect of behaviour which sometimes becomes standardized in shoots with pronounced phyllotactic specialization is the relationship between spiral directions on parent axis and lateral branch. In *Echinodorus*, for example, the last foliage leaf which the stem produces before undergoing conversion into a pseudostolon has an axillary bud of great size which proceeds at once to develop additional foliage leaves in continuation of the rosette, turning its parent axis aside as a relatively slender pseudostolon. This axillary bud, the position and function of which are highly distinctive, adopts the same direction of spiral as its parent; the situation cannot be better expressed than in Charlton's own words: 'the phyllotactic sequence of the vegetative axis continues on to the axillary bud in such a way that the bud prophyll and the pseudostolon together appear to occupy one leaf position.' The consequence of this arrangement is that each pseudostolon, although actually terminal, figures as a lateral organ in a grand spirodistichous sequence (Fig. 57) which is functionally, and in respect of its angular divergences, the equivalent of a single shoot, though it is really a sympodial assembly. As though to emphasize the uniqueness of the bud which continues the sympodium, the penultimate foliage leaf of each set (number 4 in the diagram) has an axillary bud of normal size, the phyllotactic spiral of which is reversed relative to that of the main axis.

While the phyllotactic pattern of a plant may be of interest in many respects, it is essentially a localized system of geometry, and bears no obvious or necessary relationship to the general habit of the plant or to such biologically important stages as the establishment of the seedling or the initiation of perennating organs. We have therefore to turn from the close examination of the way in which one leaf follows upon another to the broader study of changes in leaf and bud-production through the season, or indeed throughout life. It is impossible to achieve a sharp separation between the different classes of change which may be observed, but three principal considerations have to be kept in view. In many circumstances, and most conspicuously in seedlings, the quality of successive organs shows a definite sequence of heteroblastic changes; heteroblastic change is most clearly displayed in the form of the leaf, which in seedlings usually develops towards greater size and complexity node by node, but it is symptomatic of a general progression in the physiological state of the plant. Such a progression, although influenced by external conditions, must be primarily attributed to advancing age. In perennial plants we can see changes which are related less to the age of the individual than to the cycle of the seasons; the alternation between foliage-leaves and bud-scales in so many woody plants is a familiar example. Lastly we can recognize a category of more or less pure responses to environment, as in the differences between sun and shade leaves of the same species, or more dramatically in the difference between aquatic and aerial leaves of certain amphibious plants.

One of the most persistently recurrent features in studies of heteroblastic change is the tendency for lateral shoots to display some measure of reversion to the juvenile state; an axillary branch emerging from an adult shoot will not exactly reproduce the characteristic features of the early seedling stage, but will almost always display in its lower portion a measurable if not conspicuous retrogression from the adult form. A great many writers have recorded in purely descriptive terms their impressions of various heteroblastic phenomena. There is rarely any difficulty in finding means whereby the changes can be expressed quantitatively, though the appropriate form of measurement will vary according to the general morphology of the species under consideration. Thus in dealing with *Delphinium* seedlings Brown (1944) was able to proceed by simply counting the tips or points of the ultimate segments of a leaf in which the primary dissection is always threefold, with a terminal lobe and two lateral ones. In these circumstances it is possible to obtain by elementary means several different kinds of quantitative information. The first leaf has about 10 points, the tenth about 60, giving a mean rate of increase of about 5·6 points per internode. To this increase the terminal lobe make a rather greater contribution (about 2·2 points per internode) than do the lateral lobes, each of which gains about 1·7 points per internode. It can also be seen that dissection of the leaf increases more slowly than its area. From the first leaf to the tenth there is roughly a twelvefold increase in area, but only a sixfold increase in the number of points. A species having leaves of simpler outline, while possibly presenting less formidable problems of physiological interpretation, does not offer the same facilities for rapid accumulation of data. An investigator's choice of material must often represent a compromise between conflicting requirements, and in a high proportion of cases it will be necessary to resort to the use of suitably constructed ratios and indices to assess variations in the shape and degree of dissection of successive leaves. The review by Ashby (1948) will give access to further examples.

The heteroblastic sequence of changes, most conveniently studied in relation to leaf-shape but affecting every aspect of the shoot's activities, evidently gives a possible basis for a scale of development. The question arises whether we can distinguish a set of stages through which the shoot may pass, stages more or less fixed in their characteristics no matter with what speed the plant may pass through them. This is the concept of 'physiological age', or 'phasic development'. In its most literal interpretation it would mean that any modification of the environment, whether experimental or of natural occurrence, would operate primarily by setting the plant so much forward or back in a fixed programme of development; a plant grown slowly under severe conditions would resemble a younger but more favoured specimen. This is contrary to experience, and attempts to formulate the idea of physiological age in precise terms encounter insuperable difficulties. As an approximate guide in developmental studies the concept is not however entirely without merit, because it is sometimes important to recognize the broad equivalence of stages in developmental histories which differ in absolute speed rather than in any descriptive property. For example, it has more than once been suggested that in the breeding of crop plants those seedlings with the most rapid heteroblastic progression in their early foliage leaves are also generally the plants which will soonest attain reproductive maturity. This and other technological implications have attracted the attention of several Russian workers, whose ideas (though based in part on philosophies which would not ordinarily be admitted to scientific discussions in Western countries) were thought by Ashby to justify serious examination.

As regards the action of single physiological factors in modifying the sequence of developmental stages a most confused situation exists. Virtually any treatment which can be applied to a plant will have some effect. Among the investigations reviewed by Ashby were examples of leaf-shape changes related to light-intensity, photoperiod, the ratio of calcium to potassium, and the balance between nitrate and other salts.

As Ashby appreciated, the fact that such an agency is capable of affecting the course of development is no proof (indeed it is not even evidence) that the same agency operates as a normal component in the heteroblastic system. The possibility of hormonal changes within the plant must also be taken into account. It is to be expected that different factors will often operate additively; if A and B (not under strict experimental control) have already brought a shoot to the brink of some important morphological change, an experimenter who supplies C may obtain dramatic results, but these will not necessarily be repeatable with another batch of plants. It is therefore not particularly surprising, for instance, that Ashby (1950) should find the changing leaf-shape of *Ipomoea* seedlings to be greatly influenced by photoperiod, whereas Njoku (1956), in a direct continuation of the same research tradition, was unable to demonstrate any photoperiodic effect. The occurrence of such an inconsistency casts doubt upon the value of any simple observation of the influence of a single factor; some more complicated form of investigation appears to be called for.

Njoku (1956) introduced a worthwhile refinement into a study of the effect of light-intensity. In the *Ipomoea* seedling the heteroblastic sequence is one of increasing depth of lobing of the leaf. This is expressed by means of a 'shape index' which ordinarily increases from an initial value of 1 (entire first leaf) to a fairly constant 3·8 at about leaf 13. Upon growing plants in shade and in full sunlight it was found that shading greatly retards the heteroblastic rise of the index. It is a feature of this work that plants interchanged between treatments in the course of an experiment are able to make a complete adjustment to the environment into which they are moved: e.g. a shaded plant transferred to the sun as its second leaf unfolds has attained the normal sunleaf shape in its tenth leaf. The orderly nature of this recovery affords a useful check upon the basic observation and the technique of switching plants between treatments is obviously capable of further development. In a later paper (Njoku, 1956a) some ingenuity was brought to bear upon the possible existence of a 'leaf-lobing substance'. Defoliation delays the onset of lobing; does this mean that lobing is caused by a hormone produced in illuminated leaves? The form of experiment adopted (Figs. 58–60) involved three patterns of surgical operation, each designed to stimulate the growth of a lateral shoot by decapitation of the main axis of a seedling. When a plant is decapitated but allowed to retain its mature foliage leaves (Fig. 58, see p. 109) the lateral shoot develops with lobed leaves.

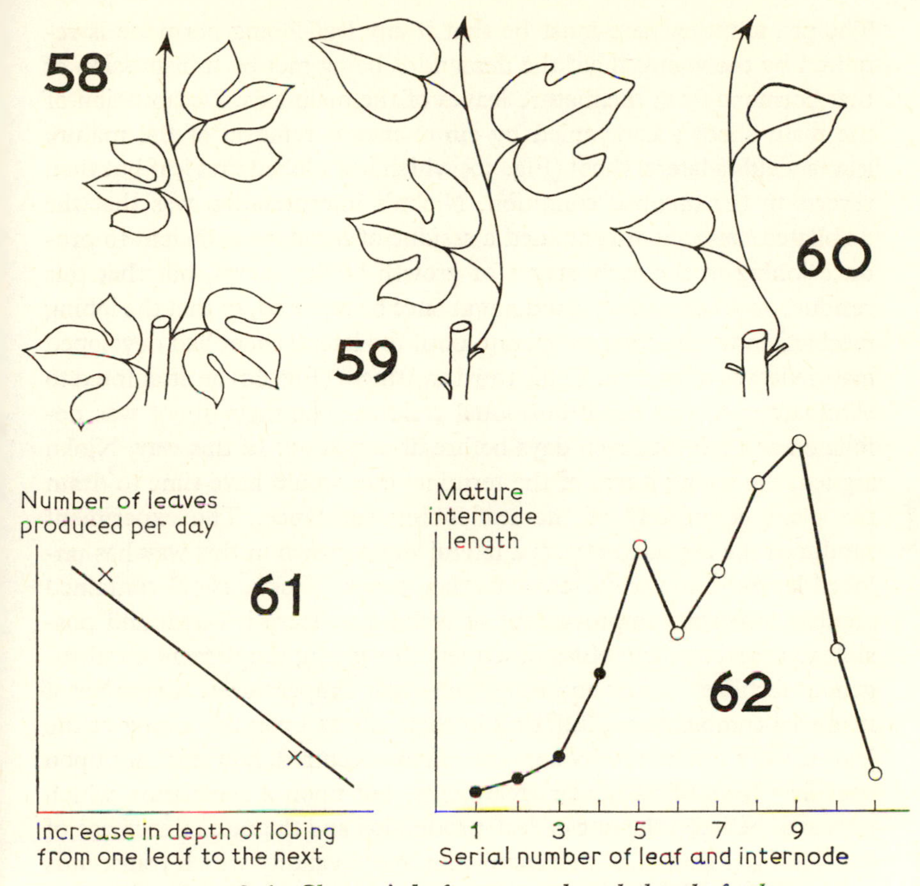

FIGURES 58–62. *Changes in leaf structure along the length of a shoot.*

58. Experiment on *Ipomoea* by Njoku (1956a); a decapitated shoot permitted to retain some mature leaves gives rise to an axillary branch which has lobed leaves from the start. **59.** Shoot decapitated and simultaneously defoliated; the branch has its first leaf lobed, presumably because of a hormone residue in the main stem, but reverts to undivided (juvenile) leaf form as that residue is exhausted. **60.** Test of the 'hormone-residue' hypothesis. Stem defoliated, then allowed to grow for a period with continuous removal of unfolding leaves, to secure exhaustion of the residue, and only then decapitated. Branch leaves undivided. **61.** Correlation observed by Njoku (1957) in the heteroblastic sequence of *Ipomoea* seedling. Rapidity of heteroblastic progression is inversely related to absolute rate of leaf production. This observation remains good even though neither of the variables can be reliably controlled by experimental treatment. Points lie scattered about the line like the two shown; it is of secondary importance that we cannot choose which part of the system will be displayed by a particular plant. **62.** Double period of internode length in a plant with foliar dimorphism of the early-leaf/late-leaf type (e.g. Critchfield, 1960). Points shown black represent internodes below early leaves.

The presumption here must be that if any leaf-lobing hormone is required by the young shoot the demand is being met by transmission of that hormone from the mature leaves of the main axis. Decapitation of the main shoot accompanied by simultaneous removal of the mature leaves yields a lateral shoot (Fig. 59) which has a lobed first leaf but then reverts to the unlobed condition. Njoku's interpretation was that the defoliated main shoot contained a residue of hormone sufficient to produce lobing in the early stages of growth of the lateral, but that this residue, once consumed, could not at once be replaced, so that the lobing mechanism would cease to operate until the lateral shoot had developed mature leaves of its own. In his third treatment (Fig. 60) he attempted to eliminate even this small hormonal residue. The main shoot was defoliated and left for seven days before decapitation. In this way, Njoku argued, the young leaves of the terminal bud would have time to drain the plant completely of the leaf-lobing substance. The experiment produced the expected effect; a lateral shoot grown in this way has unlobed leaves from the outset. A further paper (Njoku, 1957) contained another important improvement of technique. Here it was found possible to retard the heteroblastic sequence by giving supplementary nitrogenous fertilizer and by imposing high night temperatures. A number of manurial combinations, differing in their effects upon the lobing of the leaves, were systematically applied. Interest centred, however, not upon specific effects of particular treatments, but upon a correlation which appeared between the rate of leaf-production and the rate of increase of leaf-lobing (Fig. 61). Such a correlation has a validity which transcends the reliability of the means by which the plants are caused to vary. The fundamental law is a relationship between leaf-shape and the rate of leaf-development; it matters not if we cannot regulate the growth of our plants with sufficient accuracy to place a specimen at a chosen point upon the graph. A certain 'hit-or-miss' quality about our experimental procedure need not disturb the shape of the graph at all, and may indeed prove helpful in securing a properly randomized distribution of the observations.

Those fluctuations of leaf-output which are related to the passage of the seasons rather than to the age of the individual are centred about the plant's adaptation to recurrent periods during which conditions are unfavourable for growth. It is necessary always to consider the twin questions: how far, in autumn, is the behaviour of a shoot recognizably preparatory for the coming winter, and how far, in spring, is the course of development predetermined by the steps previously taken? In plants of temperate climates, and perhaps most especially in woody twigs, it is often impossible to identify any stage at which seasonal considerations can be left out of account; everything which happens is either a consequence of the previous winter or a preparation for the next. Those examples which can contribute most to a general discussion concern woody species in which there is a clear distinction between two classes of foliage leaf occasioned by the different degrees of maturity which have been attained by the component members of the winter bud. Critchfield (1960) has discussed the case of *Populus trichocarpa*, where the leaves in a winter bud can be divided quite sharply into embryonic leaves, in which the principal lateral veins are already visible, and which are at least 5 mm in length, and leaf primordia which are in a very much earlier stage of development and never longer than 1 mm. In shoots with active growth all the embryonic leaves, and some at least of the primordia, will expand as 'early' and 'late' leaves respectively. Weakly growing shoots with internodes of negligible elongation may have early leaves only, while shoots which start their growth in summer rather than in spring (exceptional in *P. trichocarpa* but a regular phenomenon, the so-called lammas shoot, in other poplars) may have late leaves only. The late leaves of Critchfield's trees were distinguished by petioles which were shorter, both absolutely and relatively, than those of the early leaves, by the much greater size of their marginal glands, and by a greater thickness of lamina which is due in the main to a greater thickness of spongy mesophyll. There is perhaps no real difference in degree of vascularization but a very conspicuous apparent difference arises on the lower surface of the fresh leaf, where the coarse texture and greater depth of the spongy tissue in the late leaves characteristically obscure the view of the finer reticulations. There are other significant differences; for instance the development of stomata on the upper surface is much more widespread in late leaves than in early ones.

Critchfield gives growth curves for leaf length on a common time-base, revealing a complex set of timing relationships. There are considerable differences between the sizes of leaves at maturity, but we shall not go far wrong if we take 8 cm as a length at which each leaf is displaying a rate of elongation very near to its maximum rate. The dates at which successive leaves attain this length of 8 cm will sufficiently show the state of affairs; the first five leaves all passed through the 8 cm stage between 9 May and 15 May, very nearly a leaf a day, and the sixth leaf (the last of the early series in that specimen) had done so by 21 May. Not until 8 June, however, did the first of the late leaves reach our standard length, and the production of five late leaves subsequent to the first was spread out over a full calendar month. Essentially the pattern is one of rapid and almost synchronous expansion of the early leaves, then a conspicuous interruption, and finally a much more leisurely sequence of late leaves. These arrangements produce a characteristic disturbance in the distribution of mature internode lengths. The internodes associated with early leaves show a progressive increase in length from the base of the shoot upwards, and the internode which separates the first late leaf from the last early one is longer still, perhaps because there is more time for it to elongate before the necessary development of vascular tissue to serve the leaf above reaches the point which makes further elongation impossible. The late leaves then usually initiate a second cycle of internode length, as shown in Fig. 62.

The heterophyllous condition of Critchfield's twigs arises through a complex disturbance of the dynamics of leaf development. In autumn leaves are passing through their early stages of growth but are then arrested, so that immature leaves accumulate; in spring the arrested growth is resumed, but with a distorted pattern of plastochron intervals. Fundamentally the whole problem appears to be one of distorted timing, and although the concepts of comparison integral and correlative plastochron have not yet been brought to bear it is clear that they are capable of such application. The disturbances of timing which are apparent in woody shoots are probably greater in magnitude than those occurring in subdecussate phyllotaxy, and are different in their arrangement, but there is no reason to regard them as an entirely separate class of phenomena; nor need we hesitate to offer the suggestion that differences of histological texture, no doubt on a modest scale, ought to be detectable among the leaves of a subdecussate system. An account of seasonal development which includes more detail about the earlier stages of leaf-growth was given by Arney (1955), working with the winter buds of *Fragaria*. The initiation of new leaves here continues throughout the winter, though at a greatly reduced rate. Depending upon the temperature the plastochron interval is about 60–90 days instead of 10–20 days as in summer. As the emergence and unfolding of leaves do not proceed during winter there is an accumulation of immature leaves in the bud; in particular the three oldest leaf initials are all at about the same stage of development, and display a 'special relationship' in their subsequent growth. In spring (Arney, 1955a) the rate of leaf initiation only very slowly increases from its winter minimum, resulting in a depletion of the stock of immature leaves. The leaf-content of the apical bud thus reaches a minimum by early May, and the season is well advanced before the early and late stages of foliar development are brought into equilibrium.

Kozlowski & Clausen (1966) have examined the distinction between early and late leaves, not with regard to its manner of origin, but with reference to its physiological consequences. They take the view that early and late leaves play very different parts in the carbohydrate nutrition of the plant. They covered leaves with foil to prevent photosynthesis and found that growth of the current year was very little affected by interference with the late leaves, but that to cover the early leaves would result in drastic stunting of the shoot, and often in its death before September. The investigation was incomplete in that no attempt was made to follow events in the subsequent season or even to observe the accumulation of reserve carbohydrates in the tissues preparatory to overwintering. The general scheme which emerges, however, is one in which reserves are expended in early-leaf production, the photosynthetic product from the early leaves then being required to sustain stem elongation and the development of late leaves and winter buds. The output from late-leaf photosynthesis, one must suspect, goes mainly into winter starch reserves, as it does not seem to be needed for any other purpose. Kozlowski & Clausen extended these considerations to *Betula* and other genera, and further pointed out that there is a significant change related to the age of a tree: long shoots with a high proportion of late leaves will be more important in young trees, whereas an old tree may have a high proportion of its leaves on short shoots which never produce late leaves at all.

The changing behaviour of woody shoots during the growing season has been studied from a different point of view by Maini (1966, 1966a) who worked with *Populus*. It appears that species can significantly differ in the form of their internode-length curve; thus *P. balsamifera* and *P. grandidentata* will both, in the course of a season, produce about twenty-seven internodes in an active shoot, and both will show an increase in length from internode to internode up to a maximum, followed by a decline. The graphs, however, are asymmetrical in different ways; the longest internode in *P. balsamifera* is about the twelfth, the longest in *P. grandidentata* about the twentieth. Maini extended his observations to the lateral buds, which also show a characteristic sequence of lengths with a maximum at some point between the beginning and the end of the year's growth. Experiments with sucker shoots of *P. tremuloides* appeared to show a direct relationship between the size of a bud and its potential for further growth. Left to itself, one of these sucker shoots will grow from its terminal bud (which is in this species larger than the uppermost lateral bud). The amount of growth so produced is not fully representative of the plant's capabilities. By removing some of the buds, or by decapitating so as to force the development of a lateral branch, it is possible to obtain a greater length of shoot in the following season than would be produced naturally. In the conditions of these experiments almost any trimming or pruning operation would result in some enhancement of the growth rate. Optimum results were however to be obtained by decapitating the sucker immediately above its longest lateral bud, the length of stem produced in this way being about twice that yielded by natural growth of the terminal bud. The effect is attributable in great measure to a prolongation of growth activity in response to the operation, and it means that even drastic injury need not be detrimental to the prospects for future development. That the relationships observed by Maini are ultimately referable to seasonal causes does not seem to be seriously in doubt; they are, however, apparently of a secondary or consequential nature, and we are very far from being able to attribute them in any direct or simple manner to specific changes in the dynamics of growth.

Even more remote from any possibility of explanation in terms of mere retardation or acceleration of developmental process are the differences in number and quality of organs which may be occasioned by changes in the nature of the environment. Many species, including probably the majority of perennials, are capable of producing two or more distinctive patterns of shoot morphology in some fairly obvious relationship to environmental conditions. Besides the ability of 'amphibious' plants to grow in 'land' or 'water' forms, which has attracted perhaps a disproportionate amount of attention, we have to consider the existence of rhizomes and stolons, the occurrence in many species of contrasting 'long' and 'short' shoots, often with distinguishable types of leaf, and many other problems. These phenomena cannot be sharply separated from those associated with advancing age, and some element of heteroblastic change will almost always be present to complicate the study of environmentally induced modifications.

The ability to produce a markedly specialized form of shoot is a taxonomic character, the distribution of which will often give an indication of the amount of evolutionary progress which has been required to give the observed end-product. Thus in a study of amphibious plants Streitberg (1954) offers comparative anatomical details of aquatic and aerial leaves, and finds that the magnitude of the anatomical difference (not always closely parallel with differences in external shape) is greater in representatives of families where hydrophytes abound, but less in those species which are exceptional amphibious members of otherwise terrestrial groups. Also, as one might expect, the greater the difference, and the more generally hydrophytic the relevant circle of taxonomic affinity, the earlier the stage of ontogeny at which the terrestrial and aquatic forms of leaf begin to be recognizable. Such relationships imply a somewhat extended evolutionary history. Where the different types of shoot in a species have been closely compared it has generally been found that they differ in a whole complex of characters, and features which are commonly regarded as reliable guides to identity are sometimes strikingly changed. For example Schwartz & Schwartz (1928), working with *Epilobium hirsutum*, in which the familiar type of shoot is erect, with opposite hairy leaves, were able by forcing the plants in a hothouse in autumn or spring, and in various other ways, to produce a creeping, freely-rooting type of shoot with almost glabrous leaves in a spiral phyllotaxy. In floristic works the inherent plasticity of species from this point of view is very inadequately indicated.

Experimental investigations designed to elucidate the causal basis for the occurrence of different shoot types in genetically uniform material have been very numerous. In reading the resulting literature it is important to maintain a sense of historical perspective and to appreciate that each generation of research workers, in choosing treatments to apply to the plants, draws preferentially upon the physiological novelties of its own time. So Njoku (1958) applied gibberellic acid to *Ipomoea* seedlings and obtained a retardation of the heteroblastic sequence. It would have been remarkable if the treatment had produced no effect. The observation, however, does no more than supplement the information which had been obtained by other methods prior to the discovery of gibberellic acid. Similarly the discovery of photoperiodism, and the widespread availability of equipment for administering photoperiodic stimulation, have made it inevitable that attempts should be made to control by photoperiodic means the balance between 'land' and 'water' types of leaf-development, etc. Such attempts have been successful in varying degrees but the results would hardly justify any claim that photoperiodic considerations should now supplant the older styles of enquiry. The truth appears to be that almost any physiologically active agent will make some contribution to the determination of the plant's behaviour and that the number of contributing factors is usually too large for any one of them to become dominant over all others.

The earlier workers were probably right in attaching some significance to the level of nutrition. Woltereck (1928) examined amphibious plants by a variety of methods, and concluded that an excess of assimilates causes a disposition to develop the land-form, whereas a relative excess of mineral ions can lead to reversion to the water-form. High light-intensity and the removal of apices (so preventing consumption of carbohydrates in new leaf-growth) may give 'land' characteristics even in submerged shoots. Warming of the soil, to stimulate root activity, combined with defoliation to cut down photosynthetic output, can yield the 'water' form of leaf even on land. Upon this view an internal change is needed but it is immaterial by what means this change is produced. The idea that aquatic plants owe some of their morphological distinction to a low rate of photosynthesis pervades much of the older literature. It appears, however, from the work of Gertrude (1937) and Gessner (1940) that the really significant feature is a lower concentration of soluble carbohydrates. The water plant is characterized not so much by a low rate of photosynthesis as by a specially prompt conversion of the products into protein and polysaccharides. This concept was given greater precision by the experiments of Allsopp (1955) on aseptic cultures of *Marsilea*. Use was here made of mannitol, a substance which appears to have no metabolic significance for the higher plants and which only very slowly enters their cells but which is osmotically active. By this technique plants can be presented with solutions of equal nutrient value but differing osmotic pressures. In the event there proved to be an interaction between the purely osmotic aspect of a culture solution and the concentration of sugar which it contains. Other things being equal, high osmotic pressure of the medium (comparable to drought in many of its effects) is conducive to the development of the land-form, yet whereas a solution based on 5% glucose produces the land-form, one of similar osmotic pressure based on 2% glucose + 3% mannitol mostly produces water-form shoots.

Other observations, however, seem not to be reconcilable with any simple and direct relationship between leaf-form and nutritive condition. For example Bergdolt (1934) worked with some of the Australian species of *Acacia*, in which the seedling has bipinnate leaves but the adult produces only phyllodes, each equivalent to the petiole and rachis of the compound leaf, without its leaflets. Reversion from the adult to the bipinnate condition could be brought about by deep shade, about 3% of full sunlight, though more readily in some species than in others and usually more readily in plants grown from seed than in those established from cuttings. This suggesting that the juvenile state represents a certain measure of carbohydrate deficiency, attempts were made, with considerable success, to induce in seedlings a precocious change to phyllode production by giving high light-intensity and supplementary supplies of gaseous carbon dioxide. This is consistent with a nutritional hypothesis, but it is not so easy to see why the removal of the pinnae from the lower leaves (presumably to the detriment of the carbohydrate level) should also provoke a premature appearance of the phyllode type of leaf.

In general the results of experimental work have been too complicated and uncertain to offer any secure basis for the explanation of field observations. It would be quite unsafe, for example, to attribute the difference between land- and water-forms of leaf to any influence of water, because it has been repeatedly shown (e.g. in *Ranunculus* by Bostrack & Millington, 1962) that one of the most potent agencies for controlling leaf-shape is temperature. The low temperature of natural waters at the relevant time of year may well be the decisive ecological factor.

Some workers have sought to identify the point in the ontogeny of the individual leaf at which the effective decision is taken to develop in one manner rather than the other. Sparks & Postlethwaite (1967, 1967a) used the heteroblastic sequence of *Cyamopsis*, in which juvenile leaves with one leaflet are followed by adult leaves with three. The difference becomes obvious at an early stage of growth in which the leaf primordium consists of a central rod with two lateral meristematic flanges. Impending subdivision into three leaflets is plainly signalled by cessation of meristematic activity in two sharply localized pockets of marginal cells. Although such an observation is valuable, it does not necessarily give the exact time of decision. In the absence of experimental evidence to the contrary it is natural to suppose that some more fundamental change must precede the first visible alteration in the cells. It is also too much to claim, as the original text shows some disposition to do, that all the differences displayed by mature leaves are merely consequential upon the choice exercised by the 'pocketal cells'.

In most of the problems which fall within the scope of this chapter it is possible to consider the growth of an individual leaf or internode as independent, in the sense that the fate of one does not automatically affect its neighbours. Cases occur, however, in which the physical conformation and texture of the parts may be such that adjoining organs are forcibly constrained into some uniformity of behaviour. In some shoot systems the parts are packed so closely together that the coordination of growth is demonstrated by an actual indentation made by one organ upon another. A case of this kind can be seen in *Saccharum* (Panje, 1961) where each leaf has a firm collar at the junction of sheath and lamina. The pressure of this upon the younger leaves inside it causes transverse 'constriction bands', which are permanent imprints upon the lamina. The print becomes successively weaker according to the number of young leaves through which it is transmitted, but the system of imprints is sufficiently well defined to show that all young leaves grow at related speeds. There is over a long period no sliding motion of one leaf through the collar of another, nor is it likely that a young leaf could accomplish such a movement without damage. Effects of this kind are probably not very rare, and they raise some difficult questions regarding the regulation of growth in the tissues.

It will be evident that many effects, whether related to age, the cycle of the seasons, the characteristics of the environment, or even to genetical differences in the plants, can be studied largely as modifications in the relative timing of various developmental processes. If we extract a correlation curve between two different variables of one leaf, as it might be lamina length against petiole length, then in passing from leaf to leaf along the shoot we shall find one of these dimensions to be somewhat advanced or retarded relative to the other, with consequent changes in comparison integral. If we plot a graph from corresponding variables of two successive leaves, as it might be length of one petiole against that of the one below, then the interval between one leaf and the next will prove to be inconstant, with consequent variation in the plastochron, whether chronologically or correlatively determined. Disturbing influences are so numerous that no shoot can in reality maintain a steady state of growth. The difficulty of selecting strictly comparable organs for purposes of physiological experiment is therefore considerable, and much greater than some investigators have supposed. In the next chapter we shall be forced to reject even the traditional belief that a leaf in a decussate phyllotaxy provides a perfect 'control' for experimental treatment applied to its partner.

No problem arises so long as it is practicable to take a complete plant as the experimental subject, or in circumstances where all that is needed is the qualitative recognition of an effect too gross in its manifestations to be obscured by morphological differences. If, however, we require a high standard of numerical accuracy in experimental work on single leaves, internodes, or lateral branches, then it is necessary to provide for the fact that the organ selected for experimental treatment will generally be unique, and that a 'control' specimen cannot be found. The appropriate response will be to develop a scheme of reciprocal comparison. To see how this would work, suppose that for experimentation upon individual leaves, we take seedlings with subdecussate phyllotaxy. For half the replicates of a treatment take leaf 3 as experimental subject, leaf 4 as the control, and in the remaining replicates vice versa. From such a system it will be possible by suitable mathematical analysis to cast out the major inherent differences between leaves 3 and 4, and to extract a purified residue of experimental effect. To choose, as a matter of deliberate policy, an experimental plant with a very large subdecussate oscillation would be foolhardy and perverse, but any idea that the need for reciprocity of comparison can be entirely avoided seems to be naïve and dangerous. In the last resort a moderate source of error, which can be clearly seen and against which systematic precautions can be taken, is infinitely preferable to a situation in which experimenters are merely hoping for the best because they have no visual warning of the complications which lie in their path.

Chapter Five

The Symmetry of the Shoot

Organisms universally display, both in structure and in function, qualities of asymmetry or one-sidedness. Everyone is familiar with examples in his own body, such as the position of the heart or the preferential use of one hand. Most human subjects, however, are quite unaware of the diversity of such effects, and are easily surprised by the simplest of demonstrations; one such trick is to ask an audience to fold arms across the chest, note how the arms are crossed, drop the arms to the sides, and then fold arms again with the cross-over reversed. In most phenomena of this type the two contrasting forms which are geometrically possible are very unequally distributed in the population, and genetical controls must be supposed to operate. The 'abnormal' state may be found in a substantial minority, as with left-handedness in man, but is often extraordinarily rare, as with reversed twist in the shells of certain molluscs. Some asymmetries, however, though deeply ingrained in the individual, are certainly acquired. English is written from left to right, Arabic the other way; the consequences, for any adult student of the 'other' language, are complex and uncomfortable, but the difficulty arises from early training, not from genetics.

Such considerations lead us to expect that every constituent organ of a shoot system will possess inherent properties of asymmetry. For example it has been known since the eighteenth century (Schmucker, 1925) that the thickening-bands of spiral xylem elements are in the overwhelming majority of cases inclined in such a way as to form a left-hand screw (according to the normal engineering convention). Claims that particular species were characterized by right-handed vessels and tracheids have never survived independent re-examination of the material, and some of their authors are known to have believed (which is untrue) that the optical system of a microscope produces an apparent reversal of the sense of a helical specimen. Right-handed vessels undoubtedly exist, but it is difficult to find an organ in which the right-handed cells form more than a tiny proportion of the whole. The distribution of such basic unsymmetrical properties among plant organs, just as among animal bodies, is of an apparently capricious and irrational nature. The lack of connection between the geometrical arrangements and any known principle of biological need is particularly clear in some of the lower plants. The chloroplast of *Spirogyra* coils to the left, the otherwise apparently similar chloroplast of *Spirotaenia* coils to the right; *Volvox globator*, at least in some strains, swims with a constant left-handed rotation, while *V. aureus* displays no such constancy but erratically reverses its direction of spin. There is nothing to be gained at present by attempting to set up Darwinian 'explanations' for this class of information.

A plant organ of any structural complexity can shelter a large number of separate unsymmetrical properties, between which there is often a striking lack of correlation. For example the generalization about the predominance of left-hand coiling in spiral xylem elements can by no means be extended to the slope of the elongated pits in the walls of xylem fibres; these cells are much more variable, and predominantly right-handed. It is impossible to avoid the conclusion that different asymmetric properties can coexist in the same tissue, and even in the same cell, with remarkably little interaction. The prevalence of left spirality in xylem elements can only be attributed to a molecular asymmetry in the cytoplasm; this molecular pattern is presumably very ancient, since every angiosperm family has inherited it, and presumably exists in every cell, since it is present in stems and roots alike and is readily transmitted through cells which are not themselves vascular. Yet this basic cytoplasmic spirality seems powerless to influence, for example, the direction of a phyllotactic spiral; in angiosperm species (apart from a few instances of genetic determination) right and left phyllotactic spirals are equally frequent, up to the statistical limitations of the available data. Other unsymmetrical phenomena, in both vegetative and reproductive morphology, display a similar pattern of independence and inconsequentiality. Whether a particular geometrical relationship which can be conceived in right- and left-handed forms will occur in only one of those forms to the exclusion of the other, or in both forms equally, or in both forms but with a preference for one, can only be ascertained by trial, and what is variable in one plant may be constant in a related species, or even in the same species at a different stage of growth.

Among the various types of spirality commonly observed in stems are the twisted grain of tree trunks and the twining of stems about a support in many climbers. From Schmucker's review of the older literature it appears that the twisting of trees is to some extent genetically determined: *Betula* and *Aesculus* twist to the right, *Castanea* to the left, while some species of *Pinus* have a left-handed twist in the young tree which changes over to right-handed with increasing age and girth. These relationships can probably be traced back to the inclination of newly formed cell-walls in the cambium, but this leaves the fundamental questions unanswered. Evidently, however, the twist of wood grain bears no simple relationship either to phyllotaxy or to the spiral patterns observed in individual cells. The behaviour of twining plants displays yet another independent system of causation, one in which questions of taxonomic affinity assume paramount importance. Setting aside a very small number of species in which the direction of twining is indifferent, so that right and left coils can occur in connected branches, or even at different levels in the same stem, it is possible to make a sharp separation between right-twining and left-twining species, with a considerable numerical preponderance of right-twining examples. The lack of balance arises because most families in which twining is a common habit of life, and in which there are many twining species, twine uniformly to the right, whereas most of the families in which left-handed twiners occur are families in which the twining habit is neither characteristic nor common. On the average, left-handed twining is indicative of a considerable measure of taxonomic isolation from other twining species. Such a situation would be expected if left-handed twining (perhaps because it was antagonistic to some inherent spirality of structure) were a less efficient procedure, and unlikely to achieve the kind of ecological status that might lead to copious further speciation.

Into the already complex asymmetry pattern of the stem the development of any lateral organ must necessarily introduce a further characteristic distortion. Something of what is involved can be seen in the work of Jahn (1941) on the epidermis of the stem in *Vicia*. Here the transverse section of an internode is quadrilateral (Fig. 63, see p. 124) and the attachment of a leaf involves three of the four corners; as the phyllotaxy is distichous the whole surface of the stem is partitioned by collenchymatous ribs into panels, each extending vertically through two internodes. Any internodal plane of section includes the upper parts of two panels and the lower parts of two others. Jahn, by extremely laborious but critical methods, determined the distribution within these panels of epidermal cell-length. In Fig. 63 the central white areas of the small charts represent transverse sections of the stem at the levels indicated by the arrows. Each face of the stem is used as a base-line for the construction of a graph (black area) showing the variation in cell-length across the panel. Thus we see that immediately below a node there is a marked shortening of the cells (i.e. an increase in the rate of transverse division) in the panels below the leaf of that node, as compared with much longer cells on the other side of the stem, and that this shortening is most pronounced at the centre of the panel. At a very slightly lower level the distribution of cell-length across these panels is quite different. The conditions at the middle of the internode are particularly significant: even at the point furthest removed from a leaf the epidermis of the stem is conspicuously and unsymmetrically distorted in relation to leaves at other levels. These findings have to be regarded, in some respects, as elementary, relating as they do to a single tissue layer and a specially simple pattern of phyllotaxy. Much more complex situations are likely to be encountered elsewhere.

It is not in the least probable that any leaf should be truly symmetrical about its central line, if only because the basic spiralities of the stem, as manifested in vessels and fibres, are known to be continued into the foliar venation. It appears however that such unsymmetrical trends as can be assigned to causes within the leaf are generally small in magnitude. Where a leaf displays asymmetry so gross as to be detected by casual inspection, the effect nearly always forms part of a larger scheme, which extends to other leaves also and bears a fixed relationship to phyllotaxy or the arrangement of branches. An example can be seen in the banana leaf (Skutch, 1927). This plant is altogether exceptional in having a constant direction of phyllotactic spiral (left-handed by engineering conventions). If we look along a leaf midrib from the stem towards the tip of the leaf, with the adaxial surface facing upwards as in in nature, then the edge of the leaf which is to our right is the anadromic edge (that is the one which points in the ascending direction of the phyllotactic spiral). In the young leaf the right half of the lamina is rolled, making at its widest part about eighteen turns, and packed into the concavity of the upper surface of the midrib. At its apical end this roll of anadromic half-lamina is finished off by a dome-shaped cap formed from the leaf margin. The other (catadromic or left) half-lamina is wrapped round the outside of this whole assembly of midrib and right half-lamina together. Because it is made up on such a thick core the catadromic roll makes only four or five turns, and it lacks an apical dome. Owing to these arrangements the direction of the phyllotactic spiral is

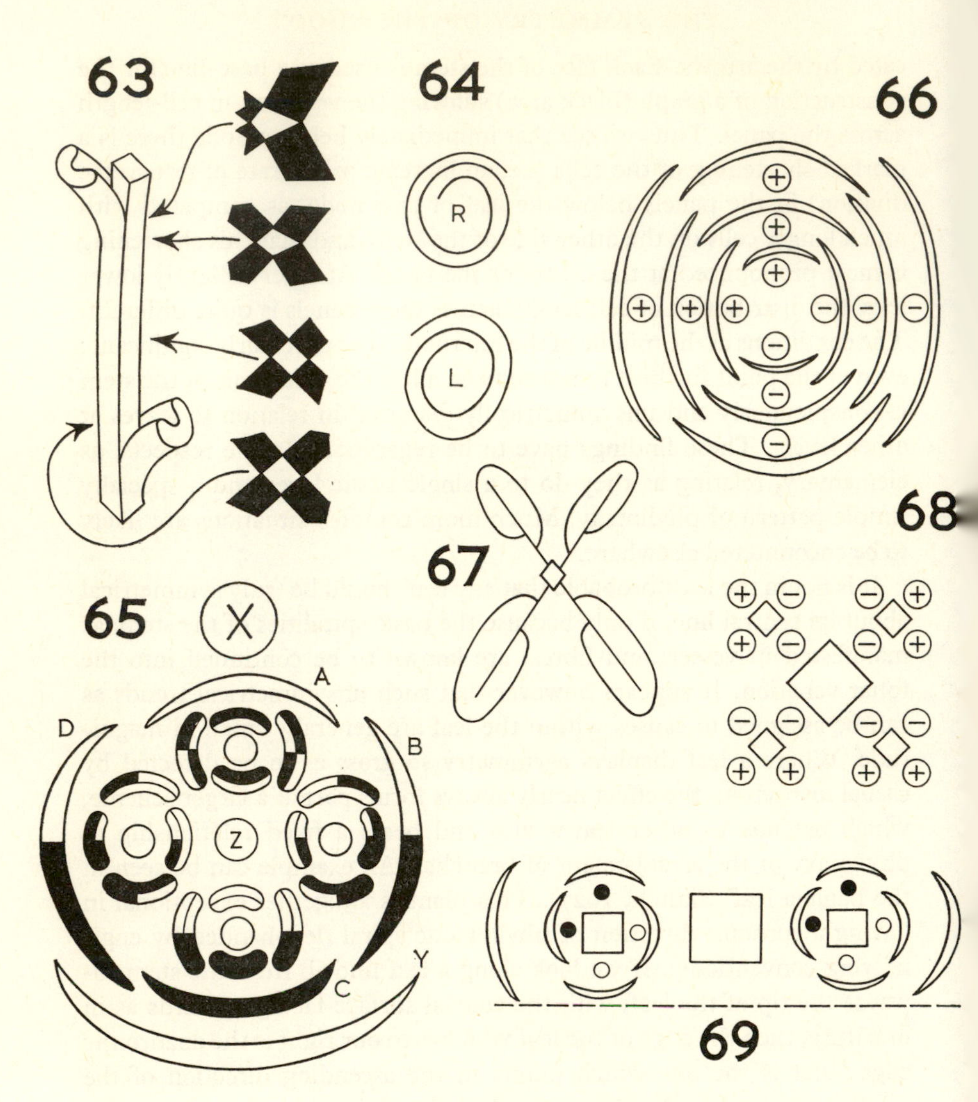

FIGURES 63–69. *Systems of symmetry.*

63. Distribution of cell-length in the internodal epidermis of *Vicia faba* (Jahn, 1941); each of the small charts represents a transverse section of the stem at the level indicated by the arrow, each side of the section serving as base-line for a graph showing how epidermal cell-length varies across the side. The phyllotaxy being distichous, the pattern will be repeated in the next internode with the charts rotated through 180°. **64.** Raunkiaer's convention for R and L rolling of the first leaf in grass seedlings, in plan view. **65.** Plan of a decussate shoot with two orders of branching; a stem X bears a branch Z in the axil of the leaf Y, and Z in turn bears four branches in the axils of leaves ABCD. Shading is applied to ABCD on the basis that parts turned towards Y shall be black. In the second-

permanently marked on the tip of every leaf. The asymmetry of the leaf in *Musa* has therefore a different status from the observation (Brett & Dormer, 1960) that leaves of *Spiraea* show a persistent bias towards the production of a slight excess of marginal serrations on the left-hand edge. This effect, or at least some part of it, may well be independent of phyllotaxy, which in *Spiraea* displays the normal inconstancy of direction.

A plant organ presumably cannot acquire an unsymmetrical property through any process of education, but it may do so through the action of other organs upon it. In a species with spiral phyllotaxy, for example, right and left spirals will generally be equally frequent, but it does not follow at all that the foliar spiral of an axillary branch will be unrelated to that of the parent stem. Raunkiaer (1919) determined for various woody species the proportion of antidromous lateral branches, that is of branches in which the phyllotactic spiral is opposite to that of the main stem. In *Salix* and *Populus* the proportion was very nearly 50%, so that homodromous shoots (in which the phyllotactic spiral agrees with that of the parent) were about equally common, but in the vascular plants as a whole there appears to be a decided bias towards the antidromous condition. The total exclusion of homodromy is perhaps uncommon, but *Crataegus monogyna* and *Sarothamnus scoparius*, with roughly 95% of their lateral branches antidromous, represent a standard of coordination which is often attained. The proportions can be changed by experimental treatment: *Prunus spinosa* has a natural level of about 60% antidromy, but this jumps to 95% upon severe pruning.

In respect of some other unsymmetrical properties which are not strictly of phyllotactic origin a similar system of imperfect correlation can be demonstrated, but with the added complication that genetical control is operative at the same time. Raunkiaer studied the rolling of the first plumular leaf in the seedlings of grasses. The lamina is here rolled edge-to-edge in a single coil (Fig. 64). This is not entirely independent of phyllotaxy, because successive leaves are rolled R L R L R . . . (or, as

---

order leaves the shading of ABCD is reproduced in the adaxial layer, but the abaxial layer is shaded with reference to the axillant leaf, A, B, C or D as the case may be. **66.** Distribution of + and − branches in various Acanthaceae (Danert, 1953). **67.** Inequalities in size and shape of leaves in horizontal shoot of *Strobilanthes*. **68.** Enlargement of the same to show ± distribution in second-order axillaries. **69.** Pair of runners in *Glechoma*, showing convention adopted by Bowes (1961) for measurement of asymmetry (see text, p. 131).

the case may be, L R L R L . . .) alternately in the distichous sequence up the stem. It is not, however, a phyllotactic spiral in the normal sense of the term. In a survey of barley varieties Raunkiaer found a constant excess of seedlings with L rolling of the first leaf (19,530 seedlings out of 33,196). This is an average proportion of about 59% L seedlings, but the proportion in single races varied from 52% to 70%. In these grasses the lamina at maturity takes on a torsion which is geometrically quite separate from the initial rolling in the bud, and which presents an almost accidental appearance; as though to emphasize the lack of any connection, the torsion of barley leaves is constantly R, by Raunkiaer's convention. The position in oats is quite different: here the rolling of the first plumular leaf is predominantly R (about 55%), but the lamina-torsion is predominantly L (about 84%). As shown in Table 13 there is a strong

TABLE 13

Relationship between rolling of the first plumular leaf in bud, and the subsequent natural torsion of its lamina, in a sample of oat seedlings. Numbers of seedlings in four subclasses from Raunkiaer (1919)

	Rolling	
Torsion	R	L
R	70	4
L	182	274

though not absolute correlation between the two effects: either direction of rolling conduces to, but does not finally determine, a torsion of the same sign. In grasses the torsion of the awn (which is perhaps morphologically homologous with the lamina of the foliage leaf) appears to be yet another more or less independent manifestation of unsymmetrical tendencies. Awn-torsion is often constant for a species (constantly L in oats), but the grasses which have been examined from this point of view fall into R and L classes (according to the twist of their awns) almost equally, and R and L species may exist in the same genus.

Where different asymmetrical configurations are associated, it is possible to calculate what would happen if the association were of a random character, and then compare this prediction with the observations. We may take for example the structure of the calyx in the flowers of *Geranium*. The sepals here show a pattern of aestivation, or overlapping, such that the third sepal in a sequence of five has two very different edges; the asymmetry may occur in either form, and enables us by agreed convention to classify each flower as R or L. Further the inflorescence in many species is of regular construction, with two flowers at the apex of a common peduncle. Some students examined the flower-pairs of a decorative garden strain with the results shown in Table 14. The calculation

TABLE 14

Correlation between calyx-aestivations of adjoining flowers in *Geranium* sp. Original observations

	Pairs with both flowers R	Mixed pairs	Pairs with both flowers L
Observed occurrences	159	197	282
Expected occurrences, if association were random	104	307	227

of the expected frequencies shown in the lower line of the table proceeds as follows:

The observed number of R flowers is $159 \times 2 + 197 = 515$, the observed number of L flowers is $282 \times 2 + 197 = 761$, and the grand total of flowers examined is consequently $515 + 761 = 1276$, agreeing with the total of 638 pairs. The inherent probability that a flower will be R is consequently 515/1276, while the probability that any two flowers not exerting any mutual influence shall both be R must be $(515/1276)^2$. The expected number of R pairs in a sample of the given size is therefore $638(515/1276)^2 = 104$. The combinations RL and LR being pooled as 'mixed pairs', the expected number for this category is $2 \times 638 \times (515/1276)(761/1276)$, which is 307. The remaining expectation may be obtained either as $638 - 307 - 104 = 227$ or as $638(761/1276)^2$. The conclusion must be that conditions in the plant are unfavourable to the appearance of mixed pairs, and indeed in other species of *Geranium* populations certainly exist in which mixed pairs are much rarer than in this example. It is not altogether satisfactory to record observations as in Table 14 because the inflorescence is so constructed that the flowers of a pair are distinguishable: each is either the right or the left member of the pair. RL and LR pairs are therefore not geometrically interchangeable, and it would be interesting to know, as at present we do not, whether the plant is equally prejudiced against both.

The principle that any asymmetry in the position of an organ relative to other parts of the plant must give rise to some asymmetry in that organ's development appears to be perfectly general in its application. When this idea is combined with the characteristic geometry of axillary branching it very soon generates a most complex pattern of unsymmetrical parts, even in the quite impossible case of a shoot system which can be deemed to start in life with no intrinsic asymmetrical properties at all. In Fig. 65 we see the plan of a decussate system which has branched twice; between the axis X and the leaf Y is the branch Z; this in turn has leaves ABCD, each with an axillary branch again with four leaves. Shading has been applied to the leaves ABCD on the basis that parts turned towards Y are to be black, those turned towards X being white. In leaves of the next order there are two asymmetries to be shown simultaneously. The X–Y axis exerts its effect on these leaves, and this is symbolized in the upper or adaxial layer of each lamina, but there is also the more local asymmetry, such as that on the Z–A axis, to be symbolized in the lower layer of lamina. Inspection of the sixteen second-order leaves will show that no two of them are identical, and that only four have even a theoretical prospect of being symmetrical about the midrib. Another order of branching, and we shall have sixty-four leaves, none interchangeable, and sixty of them irremediably lopsided in ways which do not depend on any unsymmetrical property of the apex of X. Effects which depend upon the position of origin of a shoot must be expected to decay as the growth of that shoot proceeds, and it will not be surprising if in some cases the demonstration of asymmetry demands a standard of observational accuracy which cannot easily be attained. It is inconceivable, however, that any shoot system should be truly free from this kind of geometrical complication.

Some of the possibilities associated with decussate phyllotaxy were explored by Danert (1953) in a study of the Acanthaceae. The fundamental pattern here is shown in Fig. 66, where + and – signs have been inserted in the axillary positions. These signs indicate an inequality in development, the manifestation of this inequality varying in different examples. Sometimes a + bud is simply larger or more rapid in its growth than its – partner, but more pronounced morphological distinctions can occur. Some of the Acanthaceae have an obvious shoot dimorphism comparable with that of *Pinus*, *Ginkgo*, etc.; the long shoots then arise in the + positions and the short ones, which are in some species of a spinous character, in the – positions. It is common also to find that flowers are borne only on + shoots while the – shoots are purely vegetative or may be, in the inflorescence, entirely suppressed. The existence of two types of orthostichy is in some Acanthaceae most conspicuously signalled by the leaves, those in the + ranks being larger; it is not however possible to extract from Danert's observations any regular principle of correlation, and the force of the ± distinction appears to switch from leaf to branch according to circumstances. The only real constant here is the rule that two adjoining ranks should be different from the other two; Danert reports some shoots as symmetrical, but this claim is based upon visual inspection only, and seems unlikely to survive closer scrutiny.

In most Acanthaceae the symmetry relations vary in different parts of the plant. Thus in *Sanchezia* the vegetative shoot is 'symmetrical' whereas in the terminal inflorescence only two orthostichies are fertile. *Beloperone* on the other hand, so far as its leaves are concerned, shows a conspicuous ± distinction in the vegetative shoot which is not continued into the inflorescence where the bracts of all four rows are roughly equal. These and other examples seem to rule out any possibility of associating reproductive activity simply with a general increase or decrease in shoot asymmetry. There is however some prospect of physiological or genetical elucidation in particular cases. *Strobilanthes anisophyllus* has an obvious difference in size of + and − foliage leaves, whereas *S. isophyllus*, which is unknown in the wild and is probably a cultivated variety of the other, has 'symmetrical' vegetative parts. Danert has made charts of inflorescences of both forms, showing the date on which each flower opened; from these it appears that the lapse of time between corresponding flowers in + and − situations is ten or twelve days in *S. anisophyllus* but seven days or less in *S. isophyllus*. A relationship of this kind seems wide open to various forms of experimental enquiry.

Because the two + orthostichies are adjacent the shoot has a general transverse polarity, and there is an opportunity for this to impress itself upon individual leaves and buds. In various species of *Strobilanthes* there is a distinction between vertical shoots (which are superficially 'symmetrical') and horizontal ones which show the asymmetry of leaf-shape illustrated in Fig. 67. To some extent this dimorphism can be controlled by cultural treatment; horizontal shoots revert to vertical growth on repotting, and so on. In the horizontal shoot the four ranks of leaves are placed diagonally in relation to gravity. Fig. 67 shows this orientation, the arrangement of the axillary buds being given in Fig. 68. The axillary shoots are subject to the two rules: (*a*) in the first leaf-pair the + member is the one towards the + side of the main shoot; (*b*) in the second leaf-pair the + member is the abaxial one. It is worthy of note that in the buds the + member of the first leaf-pair is turned towards the less-developed side of the axillant leaf. We have here a situation in which the behaviour of a branch may be reflecting four distinct influences: (*a*) the ± bipolarity of the parent shoot; (*b*) the abaxial/adaxial bipolarity in the bud; (*c*) the asymmetry of the axillant leaf; (*d*) the direct effect of gravity. Any scheme for separating these components must take into account the natural geometry of the system; Danert clearly appreciated, for instance, that the radial and gravitational components are reversed in their relationship as between the upper and lower rows of axillary shoots. If we apply this as an algebraic formulation, taking symbols from the list of unsymmetrical influences just given, the second leaf-pairs of + and − buds would have asymmetries involving terms in $(d - b)$ and $(d + b)$ respectively, whereas the asymmetries of the first leaf-pairs would be independent of b altogether. These differences may perhaps be detectable in the undisturbed plant, but the relationships of d with a and c could only be changed by experimental interference. It is characteristic of the capricious nature of these phenomena that the buds in these very dorsiventral shoots of *Strobilanthes* show no great difference of size, indeed the − shoot may occasionally be the larger.

The Acanthaceae have evolved in several directions involving further elaboration of their basic system of shoot construction. In some members of the family the ± distinction between orthostichies is not absolutely fixed; a shoot may for several internodes maintain + and − ranks in the normal way, but then the whole geometrical pattern is rotated through a right angle, making one of the old + rows into a −, and vice versa. Whereas some plants effect this rotation only at irregular intervals some do it at every node, giving spiral series of + and − units, still subject to the principle that there shall be one + and one − at each node. Another line of development leads to highly condensed sympodial systems. The distinctive behaviour of this family appears to be very deeply ingrained; there are suggestions for instance that the primary root of the seedling produces lateral roots preferentially upon the radii which correspond with the + orthostichies of the plumule, and also that the + side of a shoot, when cut into pieces, shows greater regenerative potential than the − side. Similar lines of specialization are found in some other families; for example Cutter (1967) has described decussate Amarantaceae with + buds in a spiral which occasionally reverses its direction.

While it is natural that special attention should have been devoted to groups like the Acanthaceae in which some of the inequalities are so gross as to impress the most casual observer, we have no reason to doubt that comparable phenomena, not necessarily upon a precisely similar plan, will be found in any plant with decussate phyllotaxy if only the observational procedures can be developed to the necessary standards of accuracy and sensitivity. The work of Bowes (1961) on *Glechoma hederacea* clearly indicates the essential form of such investigations. In this species there is no very pronounced inequality in leaf-size, but it is practicable to study quantitatively the differences in axillary branch development. Bowes was able to eliminate genetical problems by propagating all his material clonally from a single wild plant. Kept in a sufficiently warm glasshouse the stock remained purely vegetative, the shoots used being axillary runners stimulated into growth by decapitation of their parent.

The situation of two runners from opposite axils is shown in Fig. 69, which is a transverse section looking towards the apex. Bowes measured the lengths of the axillary shoots of the second order at each node of the runner. The length of an axillary shoot increases with age, and it is desirable to establish some measure of asymmetry which will be unaffected by this elongation. Bowes therefore resorted to percentages; his explanation is confused but he eventually arrived at a simple and effective device. At any node the mean length of the two opposite axillary shoots is first calculated and then the length of a selected member of the pair is expressed as being more or less than the average by a certain percentage. Thus if a node has axillary branches measuring 12 cm and 8 cm respectively the mean is 10 cm and the longer shoot will be recorded as + 20%. In dealing with a pair of runners as shown in Fig. 69 the asymmetry of a runner node is taken to be the percentage of the left-hand branch for even-numbered nodes, or the upper one for odd-numbered nodes, that is to say, of the branches shown black in the figure. In the course of what seems to have been intended merely as a preliminary inspection of a small number of specimens, Bowes became convinced that the asymmetry of each node, expressed as a percentage, increased rather steeply as growth proceeded. The evidence for this was really very slight, and it is a matter for regret that he did not persevere with the application of the percentage method. His presentation includes a graph showing separately the growth of the longer and shorter axillaries at the sixth node of a runner; it appears from these observations that the percentage asymmetry of that node did not greatly change during a more than fivefold increase in the lengths of both axillary shoots, and in any case such graphs would very readily furnish any correction-factor which might prove to be necessary. As it is, however, the potentialities of the percentage system can be illustrated only from some observations which were certainly too few in number to establish any significant trend. In a pair of runners, taking the first ten nodes of each, the average asymmetries were: odd nodes of left runner + 18%, even nodes of left runner + 83%, odd nodes of right runner + 9%, even nodes of right runner − 72%. From the results for even nodes there appears to be a strong tendency for the axillaries which are turned towards the axillant leaf of a runner to be the longer ones of each pair. For the odd nodes the occurrence of positive values would suggest, if we could give credence to anything based on so small a number of measurements, a gravitational effect, the upper axillaries being longer. But what would be the implication if the minor differences between the four percentages were to be repeated in further batches of material? So far as the difference between 9% and 18% is concerned, it seems that we would have to reckon with a spiral trend in the parent shoot, a circular morphogenetic field analogous to the circular magnetic field surrounding an electrical conductor. Such a field would act in concert with gravity at the odd nodes of one runner, in opposition to it at the odd nodes of the other ($9 = a - b$, $18 = a + b$). The difference between 72% and 83% would involve another phenomenon altogether, presumably an interaction between radial and gravitational forces depending on the direction in which they are crossed. From the theoretical point of view it is of the highest importance to determine the complexity of the system under examination. We have already a great number of examples of the kind of asymmetry which Bowes and Danert were examining, and for the future one may reasonably question the value of any enquiry which stops short of a critical test for the types of interaction which as yet remain so tantalisingly uncertain. There is a clear need also for experimental verification, at a more rigorous standard than has yet been reached, of any influence attributed to gravity. Bowes, for instance, took the trouble to constrain some runners into vertical growth, but his account of the results is essentially descriptive and his technique was open to objection because it involved defoliation; like some other comparable studies, this leaves the fundamental issues largely untouched.

A system of symmetry which appears in horizontal twigs of *Acer saccharum* was measured by Sinnott & Durham (1923) who made all their observations upon a single tree. The pattern is shown in Fig. 70 (see p. 134), where the sizes of the leaves are best represented by their area, while differences of shape are indicated both by the measured lengths of the two chief lateral veins of each leaf and by two ratios, A being the quotient (blade length)/(petiole length) while B is (blade width)/(blade length). The leaves in the median plane are almost perfectly symmetrical, as is shown by the lengths of their lateral veins, but the upper one is much smaller than the lower, has a lamina which is much broader in proportion to its length, and possesses a petiole which is shorter both absolutely and relatively than that of the opposite leaf. The leaves in lateral positions are in all respects of intermediate size and proportion, but are in addition of markedly unsymmetrical form; the lengths of their lateral veins indicate a difference between the gravitationally upper and lower parts of these leaves which is entirely concordant with the relationship between the other two orthostichies. The authors of this diagram, having regard to experiments in which comparable systems of symmetry in woody plants have undergone more or less complete reversal after the shoot has been forcibly displaced through 180° about its longitudinal axis, regarded the observed inequalities as a response to gravity. The physiological literature upon this matter seems to establish that the asymmetry of horizontal shoots in woody plants is very generally controlled by the position of the axis and that inversion of the shoot will produce, usually after appreciable delay, an approximate reversal of the whole system of morphological differences. That this is entirely a gravitational effect is not quite so clear, the experimental procedures which have been employed not being always sufficiently precise to exclude considerations of unilateral illumination, etc.

In a shoot which continues to grow horizontally any system of dorsiventral arrangement which results from the horizontal posture is likely to be persistent, whereas an asymmetry which is merely impressed upon a shoot at its origin may well be transitory. This aspect of the matter has been considered by White (1955, 1957), using seedlings of *Acer pseudoplatanus*. Although his conclusions rest upon a very slender basis of measurement, there seems to be no reason to doubt the validity of his

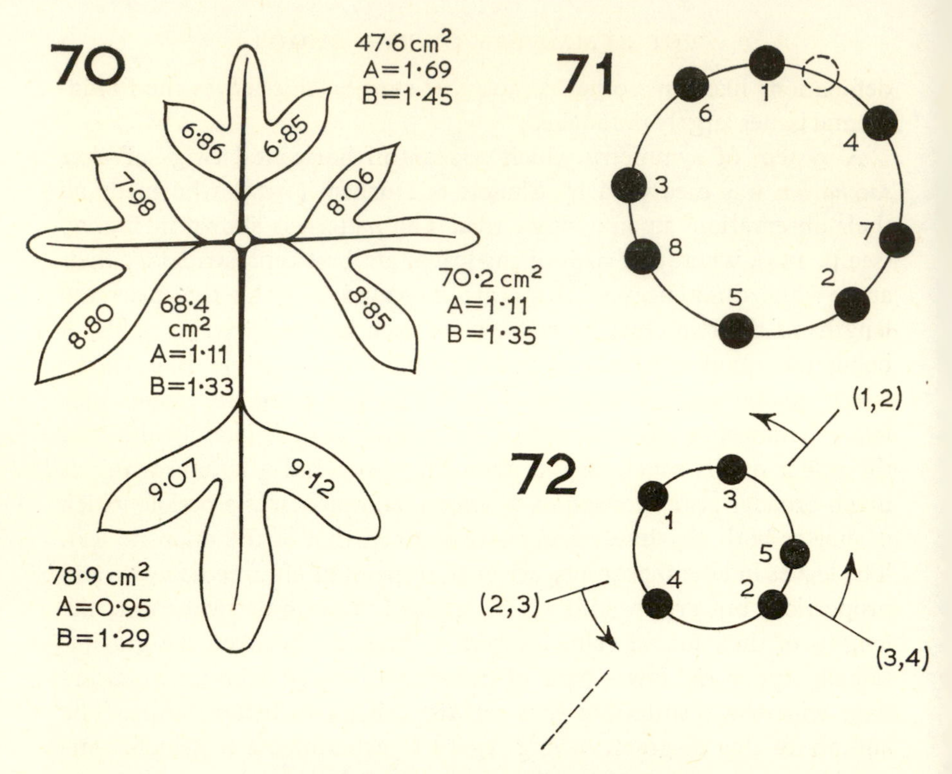

FIGURES 70–72. *Symmetry relationships.*

70. Impression (not to scale) of gravitationally induced asymmetry in horizontal shoots of *Acer* (Sinnott & Durham, 1923). The number on each lateral leaf lobe is the length of its main vein. The other quantities attached are leaf area, and two shape-indices, of which A is the quotient (blade length)/(petiole length) and B the quotient (blade width)/(blade length). All measurements vary systematically from top to bottom. **71.** A ring of eight parts in Fibonacci divergence. There are three minor intervals of about $32\frac{1}{2}°$ and five major ones of about $52\frac{1}{2}°$. The removal of number 8 would eliminate a minor interval and create a new interval of about 85°; to establish by further subtraction another approximate radial symmetry would require the elimination of the remaining minors by removal of 7 and 6. Addition of number 9 at the position indicated would divide a major interval unequally (roughly $32\frac{1}{2}°$ and 20°); restoration of symmetry by addition would then require every major interval to be similarly divided, by addition of parts 10–13. **72.** Schmucker's (1933) construction for decussate-to-spiral transition. Parts 1 and 2 are diametrically opposite. Every other part is located by rules: (*a*) bisect the larger angle between the two preceding parts; (*b*) from the specified bisector displace the new part through a constant angle α taken in whatever direction moves 3 towards 1. Each bisector has been labelled with its origin, and α has been made 26° anticlockwise, giving a rapid approach to Fibonacci conditions. The instructions contain an ambiguity regarding the position of 3. With two angles of 180° equally eligible for bisection we might have used the opposite (broken line) radius to the bisector (1, 2) actually chosen; α would then have been taken clockwise. Once 3 has been placed, no ambiguity remains.

distinction between 'primary anisophylly' (due to axillary origin) and 'secondary anisophylly' (sustained by gravity or other environmental factors). Using the same percentage notation as Bowes, White found the inequality at successive nodes of the erect plumular shoot to fluctuate irregularly between 5% and 15%. An axillary shoot begins life with a regular oscillation, about 5% to 10% inequality at odd-numbered nodes, from 45% to 80% at even ones. When an axillary shoot is caused, by decapitation of the main axis, to grow vertically, the dorsiventrality decays, and is inconspicuous (measurements not given) in the second season. If the direction of growth continues to be horizontal, the dorsiventrality persists.

Every leaf, shoot, or shoot system evidently has to be regarded as a field of action within which various systems of inequality will have their effects. These effects are however by no means unlimited. The symmetry of plants is imperfect but it is quite sufficient to assure us that the forces of asymmetry, whether the causes which bring them into play are internal or external, meet almost everywhere a significant measure of resistance. This resistance has a purely quantitative aspect which is as yet hardly accessible to experimental enquiry. To ask what factor prevents a shoot which has become anisophyllous under gravitational stimulus or as a result of axillary origin from exhibiting an even greater degree of response may at some future date give rise to fruitful investigations; that any constructive development could be founded on such a question now is far from obvious. Symmetry patterns cannot however be treated entirely in quantitative terms. An important element in the plant's resistance to disturbance lies in the tenacity with which particular geometrical configurations are replicated. A decussate shoot system, for instance, may under various influences display complex patterns of anisophylly, but it will not necessarily or even usually change to spiral phyllotaxy or begin to produce its leaves in whorls of three. In respect of these geometrical properties a shoot will ordinarily show an inertia or inherent stability which is much greater than that pertaining to purely quantitative features such as a ratio between the sizes of two leaves. This geometrical type of inertia, in which there are many components other than phyllotaxy, can in some species be overcome, with varying frequencies and with varying degrees of experimental repeatability. The resulting changes constitute a difficult but not impossible study, from which we may extract some rather limited conclusions about the nature of the conflict between the forces of asymmetry on the one hand and the fundamentally radial/concentric geometry of apical growth on the other.

An extensive literature, into which we hardly need to enter, has been concerned with the relationship between spiral and verticillate phyllotaxy, and with the recognition in 'false' or imperfect whorls of the vestiges of ancestral Fibonacci sequences. In this connection it is to be observed that (the Fibonacci angle being an irrational fraction of a circle) a Fibonacci sequence would possess perfect radial symmetry only if it contained an infinite number of leaves and if all the age-differences between those leaves were zero. In such a system any radius whatsoever would pass through one leaf and no more, but it is a basic property of the Fibonacci fraction that with any finite number of leaves in the sequence the radii which they occupy will be unequally spaced. To explore the implications of this it is only necessary to evaluate the Fibonacci fraction to a sufficient number of decimals and then to multiply it by successive integers: 1, 2, 3, etc. From each product discard any whole number, keeping only the fractional portion (because rotation through a complete circle has no effect on the angles), and arrange the fractions in numerical order. Finally subtract to obtain the intervals between them. Investigation along these lines will show that the residual imperfection of the radial symmetry is always at a minimum when the number of parts in the sequence is a Fibonacci number. Fig. 71 offers a visual presentation of this relationship for a set of eight, as more fully explained in the caption.

We have little reason to doubt that the principle exemplified in Fig. 71 lies behind the widespread recurrence of Fibonacci numbers in involucres, in respect of various other specialized structures such as the ray florets of Compositae, and in many aspects of floral construction. It is necessary therefore to be prepared for complications wherever a shoot has a rather small number of leaves and retains some impression of a spiral phyllotaxy. Some of the complicating factors are of phylogenetic origin. If for example a composite species which typically possesses eight ray florets comes under evolutionary pressure to increase that number it is not likely to proceed on Darwinian principles of gradual change. A more realistic prediction is that the capitula will become dimorphic, with alternative symmetries of eight and thirteen, and that this condition may persist long enough to permit some independent genetical adjustment of the two courses of inflorescence development, so that two distinct categories of ray floret begin to emerge. The whole effect of Fibonacci considerations is to emphasize and sharpen the differences in status between lateral organs. The ninth bract in an involucre of 9, for instance, will often be distinguishable as an intruder, uniquely disfavoured in matters of physiological competition, and peculiarly liable to disorders of development.

Increasing equality between the members of a whorl will come as their development becomes more nearly simultaneous and the rigidity of the Fibonacci scheme of angular distribution is relaxed. In many cases the angular displacements involved in this process of equalization will be very moderate. The complete freedom of any system of verticillate phyllotaxy from residual Fibonacci components can never be more than a matter of conjecture, but it is not difficult to find examples in which one can assert with some confidence that the Fibonacci element must be extremely small. *Equisetum* occupies a unique position as the survivor of the only major series of vascular plants in which the Fibonacci principle seems never to have established itself. Other systems which appear to be purely verticillate are found in such angiosperms as *Elodea*, *Casuarina*, and *Hippuris*. On the other hand there are many shoots which give a clear visual impression of whorls but which obviously contain active remnants of a spiral system. Such are the imperfect whorls of species of *Lilium* and *Fritillaria* and of various exceptional dicotyledons such as *Acacia verticillata* (Dormer, 1944). The assessment of these matters must depend to some extent on the degree of taxonomic isolation which surrounds a particular verticillate example, and can sometimes be checked by reference to special points of morphological and anatomical adaptation. Thus we may be disposed to admit the purely decussate quality of the phyllotaxy in Labiatae, where decussation is universal and associated with a distinctive modification of vascular structure (see p. 167), but feel less confident about the absence of spiral trends in Schrophulariaceae, where neither of these conditions is satisfied.

In species with verticillate phyllotaxy shoots are sometimes found which have an 'abnormal' number of leaves in each whorl, or which have (less commonly) broken away from the verticillate pattern altogether, and produced leaves in a spiral sequence. In all such cases the unusual type of phyllotaxy tends to establish itself as a stable condition in the growth of one individual stem, either from the beginning or by some process of transition as development proceeds. It is common, however, for the lateral branches of the affected shoot to revert to the normal pattern, and the extensive literature on these phenomena is dominated by the consideration that the observed effects are not under complete genetical control and are not reliably transmitted even in clonal cultures;

genetical differences do of course arise in respect of the frequency of occurrence of the abnormalities. Variation in the number of cotyledons in the seedling appears to be governed by similar principles.

The appearance of spiral phyllotaxy in *Fraxinus*, which is normally decussate, was studied by Schmucker (1933). The two leaves of a pair may in various ways show a difference in their development. Sometimes the peculiarity is no more than the appearance of a short internode between the leaves; with or without this distinction of level it is also possible to have differences in size or shape. In the lower part of the terminal overwintering bud, for instance, one leaf of a pair may be completely reduced to a scale while its partner retains the characteristics of a foliage leaf to a greater degree. All such inequalities tend to be repeated by later leaves. In most cases the disturbance ultimately dies away and the decussate condition is re-established. Sometimes however the pairing of leaves is entirely broken down, and a regular spiral pattern emerges. Schmucker found cases where spiral behaviour had persisted through the growth of at least four successive seasons, but only the main apex was affected; as is usual in these matters the lateral branches were normal decussate shoots. Schmucker discussed the angular displacements of leaves which would be involved in a transition from decussate to spiral phyllotaxy and introduced the useful concept of an angle α, which is to be the angle by which the midline of any leaf is deflected from the bisector of the larger angle between the two preceding leaves (Fig. 72). Repeated application of this principle leads to an interesting result: the divergence angle between successive leaves oscillates with diminishing amplitude about a final equilibrium at $(120 + 2\alpha/3)^\circ$. To attain a Fibonacci spiral the required value of α is about $26\frac{1}{4}^\circ$, which appears to agree with observation.

The situation regarding the occurrence of whorls with abnormal numbers was reviewed by Sitte (1957), whose extensive table makes it quite clear that by far the commonest case is that of the normally decussate species which occasionally produces whorls of more than two. To some extent, however, the records are falsified by psychological factors. In species with many-membered whorls there is rarely a definite 'normal' condition, but rather a range of variation which passes without comment; the difference between 6 and 8 is perhaps no less important than that between 2 and 3, but it makes a much less immediate visual impression. It is evident also that the casual observations which Sitte has assembled from various sources are very unevenly distributed as between species. He rightly comments on the fact that even a rare abnormality in a common labiate stands a good chance of being reported while, by contrast, the absence of any such report for an inconspicuous species like *Linum catharticum* is not to be taken as decisive. Sitte's own more intensive searches, involving in some cases the mapping of the distribution of abnormal specimens, tend to show that many species, normally decussate, can produce shoots with reasonable stable whorls of three or more leaves, but that the frequency of such occurrences is usually less than 1% and that the affected shoots are sporadically distributed, with no obvious reference to ecological factors. Only two expedients appear to be generally effective in increasing the output of pleiomerous shoots. Improvements in nutrition can have a marked influence: thus Braun (1957) found that the plants of *Epilobium*, within certain limits, produced more three- and four-membered whorls if given fertilizer and larger pots. To cut the plants down to ground level is however usually the most advantageous course; the stool shoots which then spring up from the stumps have in many cases a greatly increased propensity towards pleiomery. In this way Sitte was able to increase the proportion of pleiomerous shoots in *Fraxinus* from ½% to 3% and in *Lonicera* from 5% to 60%. There is no evidence that such manipulations can do more than encourage tendencies already existing in the plant.

Sitte attempted quantitative comparisons between shoots of the same species with two- and three-membered whorls. His measurements of shoot diameter show clearly (contrary to what has often been asserted) that in some species there is no obvious relationship between phyllotaxy and the size of the apex. Decussate and trimerous shoots have similar ranges of apical size, suggesting that the effect of nutrition upon phyllotaxy must be largely independent of considerations of general vigour. It might be expected that leaves would tend to be narrower as a whorl contained more of them. By strict proportion the breadth/length ratio for leaves in whorls of three should be 2/3 of that for decussate leaves of the same species. From observations on nine species, Sitte obtained fractions ranging from 0·872 to 1·088, suggesting, as direct examination of apices tended to confirm, that the insertion of the additional leaf is associated much less with any change in shape of the primordia than with a reduction in the spaces between them. In some instances the narrowing of the intervening strips of stem surface consequent upon the presence of a third leaf is plainly visible in adult nodes. Sitte also raised, without himself being able to reach any firm conclusion, the question of the residual inequalities in whorls of different constitution. Whether the number of leaves in a whorl has any significant influence on the extent to which an individual leaf may differ in size from its partners remains as yet in doubt.

Braun, in his work on pleiomerous species of *Epilobium* and in some less extensive related studies on *Lysimachia*, found that the plumular shoot in a range of species is normally decussate. Even in the tricotyledonous seedlings which occasionally occur the plumule does not invariably continue the trimerous symmetry but may revert to the decussate condition. It is in lateral shoots, but not usually in all of them, and not necessarily from the very beginning of their growth, that the whorled adult condition of these plants usually first appears. In hybrids between *Epilobium trigonum* and other species the tendency to produce whorls of three or four appeared to be dominant, but the expression of the character was very inconstant.

There must in fact be a serious doubt as to whether systems of verticillate phyllotaxy, even those in which there is no detectable Fibonacci component, will afford the best class of object for the study of that aspect of phyllotaxy which is concerned with the equal circumferential distribution of material or of developmental potential. In this respect much of the earlier literature is certainly defective, presenting an impression of geometrical regularity which closer observation will not sustain. The investigation of *Equisetum* by Bierhorst (1959) showed very clearly that the stability of a verticillate system is essentially dynamic: equality of number between successive whorls is hardly ever maintained over any great length of stem, and wherever there is a discrepancy of number between any whorl and its predecessor consequential inequalities of size appear at both nodes, with significantly wider or narrower leaves in specific positions. Bierhorst was able both to measure these inequalities and to extract some part of the code of law which governs their distribution; he was also able to demonstrate significant but imperfect correlations between the phyllotactic disturbances and the internal vascularization of the stem. The general conclusion appears to be, that pure verticillate phyllotaxy is not in reality of simpler constitution than a Fibonacci sequence. Inequalities between the parts are not necessarily smaller in amount; they are merely less predictable, being related to a great variety of small and local causes instead of springing from a coherent scheme of geometry. The work of McCully & Dale already quoted (p. 49) represents a level of explanation beyond which it may be difficult to make much further progress.

Apart from phyllotaxy the most conspicuous manifestation of the purely radial aspect of shoot symmetry lies in the division of the internodal vascular tissue in many cases, though by no means all, into a determinate number of vascular strands. It will become clear in the next chapter that species differ greatly in the rigidity with which their vascular symmetry is related to the phyllotactic pattern. In some cases there is an absolute, in others a merely statistical correlation, but it is clearly impossible to attribute to the vascular symmetry of most internodes any really significant degree of independence. Many investigators have however appreciated that the hypocotyl and epicotyl of a seedling constitute an area in which the phyllotactic geometry is unlikely to have assumed its normal measure of dominance, and have further perceived that the widespread occurrence of variability in cotyledon-number provides a unique opportunity to study the responses of a radial symmetry system.

The largest enquiry is that of Harris, Sinnott, Pennypacker & Durham (1921, 1921a, 1921b, 1921c). They used *Phaseolus vulgaris*; as this is normally inbred it was possible to make comparisons between several pure lines and so obtain information about genetic variation in the response to cotyledon number. Fig. 73 (see p. 142) represents an idealized construction for a dimerous seedling, in which there are two cotyledons, and two leaves (of course at right angles to the cotyledonary plane) at the first plumular node. The vascular arrangements are shown in a series of transverse sections to the right of the general view. The radicle is tetrarch, and upon entering the base of the hypocotyl the root poles divide, giving rise, in the terminology of the American workers, to four 'primary double bundles'. Through most of the length of the hypocotyl the association in pairs can no longer be recognized and there are eight equidistant strands. On approaching the cotyledonary node these recombine in pairs (this time diagonally to the root poles) and each of the resulting complexes branches into three, giving twelve bundles in all. Each cotyledon takes two of these and four others fuse in pairs, leaving six to enter the base of the epicotyl (the internode above the cotyledons). As each of these forks into two, the epicotyl ought to have twelve for most of its length. A trimerous seedling (that is, one with three cotyledons and a whorl of three leaves at the top of the epicotyl) would be expected in principle to show at every level a 50% increase in the number

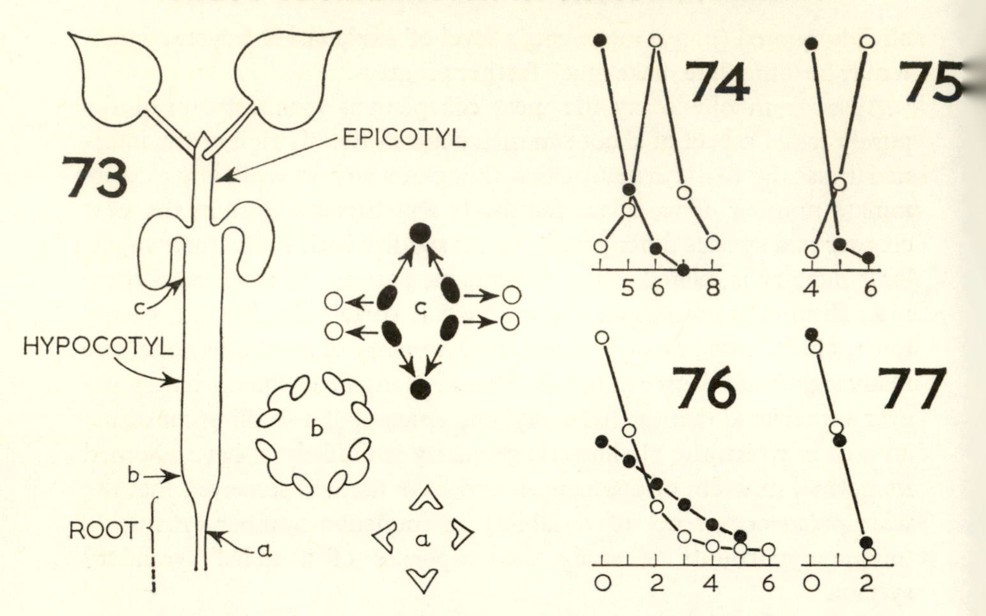

FIGURES 73–77. *Vascular symmetry in* Phaseolus *seedlings* (Harris, Sinnott, Pennypacker & Durham, 1921).

73. Schematic view of an idealized dimerous plant, *abc* being transverse sections of the vascular system at different levels with the cotyledonary radii directed to the sides of the page. Section *a* shows the four xylem poles of the root; in the hypocotyl (*b*) each pole has bifurcated and the shanks are moving towards a new association indicated by the linkages, creating four strands diagonal to the cotyledons. In the cotyledonary node (*c*) each of the units derived from *b* divides into three as shown by arrows, the white strands entering the cotyledons. The epicotyl has twelve bundles, two from each black strand of *c*. **74–77**, generalized frequency distributions for counts of vascular bundles in hypocotyls, black symbols referring to dimerous seedlings, white symbols to trimerous ones. **74.** Primary double bundles in dimerous and trimerous individuals of a race with normal variation. **75.** Corresponding distributions in a race with a genetic restriction of variability. **76 & 77.** Counts of intercalary bundles in the genetical conditions of Figs. **74** & **75** respectively.

of strands, with a hexarch root, twelve bundles at mid-hypocotyl, and eighteen in the epicotyl.

Harris and his collaborators sectioned large numbers of seedlings, both dimerous and trimerous from several pure lines, and found that the variation which actually occurs will not fit into any such simple scheme, however convenient and necessary it may be to base one's ideas upon a regular pattern. They found that the primary root of the trimerous plant, though more variable than that of the dimerous form, is markedly reluctant to increase the number of xylem poles in full proportion. Hexarchy is much less common than one might expect, and tetrarchy sets in at a lower level even in plants which have more than four poles in the upper part of the root. In the base of the hypocotyl the 'primary double bundles' are in reality very often supplemented by additional or 'intercalary' bundles. The two types can be recorded separately, giving formulations such as (4) + 2, meaning 4 double bundles and 2 intercalaries. The constitution of the hypocotyl base ranges in dimerous seedlings from (4) up to (6) + 1 and (7), and in trimerous ones from (4) up to (8) + 1. Thus, although dimerous seedlings are most commonly (4), or possibly in some lines (4) + 1, while trimerous ones are most commonly (6), the variation is thoroughly transgressive; the sight of a hypocotyl section will not generally afford a clear indication of the number of cotyledons. Despite the unexpected flexibility of its organization, however, the hypocotyl is more responsive than the root to the addition of a cotyledon. Trimerous plants have six primary doubles in the hypocotyl much more often than they have six poles in the radicle.

The question of racial differences assumes in the hypocotyl base a disturbingly complex form. Taking first the frequency distributions for counts of primary doubles, we find in most lines the situation shown in Fig. 74, where the trimerous population is spread almost symmetrically about a mode of (6), while the genetically identical dimerous plants can deviate only in one direction from their mode of (4). Line 139, however, was distinguished by an inherent veto on numbers in excess of (6), producing the graphs of Fig. 75. The statistical and genetical behaviour of the intercalary bundle counts is quite different. The modal number of intercalaries being 0 throughout, and the average number of intercalaries per plant being always less than 1, the distributions are always skewed. There appear to be two main genetical possibilities. Some races, as in Fig. 76, have a rather wide range of variation and a significant difference in form of curve between the dimerous and trimerous components of the population. In others (including line 139) the numbers are less variable and the graphs for dimerous and trimerous plants are similar (Fig. 77). The evaluation of genetic differences seems therefore to call for the use of such quantities as: (percentage of trimerous seedlings having no intercalary strands) *minus* (percentage of dimerous seedlings having no intercalary strands). This number had values from 20 to 41 in three of the strains examined, and was virtually zero in two others.

One might expect the number of bundles in the middle of the

hypocotyl to be: (number of intercalary strands) *plus* 2 × (number of primary doubles). In reality it is possible for intercalary strands to appear at the higher level without any downward continuation, while the division of the primaries is not always regular. The proposed calculation (though usually giving a correct result) is thus subject to a margin of uncertainty, and the breadth of that margin is different in different stocks. There is an opportunity here for genes which do not operate on the hypocotyl base to exert a modifying influence. A further complication is the existence, at least in trimerous plants, of a negative correlation between intercalaries and primary doubles, implying some degree of physiological stabilization of total bundle number.

The examination of epicotyls tends to emphasize the importance and completeness in *Phaseolus* of the vascular rearrangement which takes place at the cotyledonary node. The vascularization of the epicotyl, though not quite indifferent to cotyledon number, affords a most imperfect guide to seedling symmetry, the variation which exists in the epicotyl being largely independent of arrangements at the lower levels, while genetic differences which have an obvious effect upon the hypocotyl have little or no influence above the cotyledons. The general picture which emerges from this major piece of research is one of contrast between the embryo, a remarkably plastic organism in which gross structural irregularities are commonplace, and the plumular shoot, which progresses very rapidly towards effective standardization of its own internal arrangement.

By simple observational methods it appears that we may with some confidence resolve the general problem of symmetry into three components. There is firstly the type of unsymmetrical property which is intrinsic to a particular organ and which must be attributed to genetical causes or to pure chance or to some combination of the two. Secondly we have the unilateral forces of more or less obvious origin, arising from external agents such as gravity or from geometrical relationships like those involved in axillary branching. Lastly, there is the background of stabilizing and equalizing mechanisms tending to maintain a radial pattern in a stem or shoot, a bilateral one in a leaf; this background is an elusive subject for study because it can rarely be seen in anything approaching a pure state, because its effects are largely of a geometrical character, and because its principal manifestation is an array of negative results, a persistent failure to obtain statistically significant differences. For the more detailed analysis of symmetry situations two lines of action present themselves: we may resort to experiment, or we may develop more refined techniques for the quantitative analysis of observational data.

In the experimental field the possibilities are obviously very varied: we may apply treatments in a settled geometrical relationship to the morphological configurations, as by surgical removal of half a leaf or by unilateral application of a hormone, or we may observe the effects upon symmetry systems of physiological treatments, such as a change in day-length, which do not in themselves contain any specifically directional feature. Existing literature is characterized by a heavy emphasis on surgical experiments, more particularly in relation to phyllotaxy, and on localized hormone applications often running closely parallel with surgical procedures. We may review a few examples without attempting to follow this fashionable trend.

The anisophylly of *Coleus*, in which a lateral shoot normally produces larger leaves on its abaxial side than on the adaxial one, was investigated by von Guttenberg & Müller (1957), who worked largely with stem-cuttings which had a single node and which were taken from a main axis so as to ensure that the two leaves at that node should be of equal status. These cuttings were taken variously from old and from young stems, and while some of them were allowed a free development of roots others were subjected to a regimen of root-pruning. The degree of anisophylly was conveniently expressed as the ratio of length of the lowest abaxial leaf of an axillary shoot to the length of its adaxial partner. Table 15 shows

TABLE 15

Measurements of asymmetry in axillary shoots from *Coleus* cuttings (von Guttenberg & Müller, 1957). Four types of cutting, three styles of surgical preparation (Figs. 78–80). Entries are quotients: (lengths of first abaxial leaf)/(length of first adaxial leaf)

	Parts removed			
	None	One leaf		Leaf and bud
Type of cutting		Leaf present	Leaf absent	
Old rooted	1·68	1·91	1·35	1·24
Young rooted	1·60	1·99	1·42	1·16
Old root pruned	1·55	1·72	1·46	1·27
Young root pruned	1·42	1·32	1·48	1·43

the results obtained in three experiments, the surgical procedures of which are illustrated in Figs. 78–80 (see p. 147). The interpretation of the first three lines of this table offers no particular difficulty. In the first column we see an approximately constant degree of anisophylly arising in the (equal) axillary shoots of a symmetrical cutting. In the next two columns we see that in a cutting from which one of the original leaves has been removed the axillary shoot of that leaf is less unsymmetrical, but the axillary shoot of the surviving leaf considerably more so, than would have been the case without surgery. This is readily intelligible in terms of mutual inhibition. An axillary shoot is inhibited by its own axillant leaf. Removal of a leaf therefore liberates its bud, which grows larger than that of the surviving leaf. But this very active shoot will itself exert an inhibition upon its less energetic neighbour, and this inhibition will differentially affect the nearer (i.e. the adaxial) leaves of the smaller shoot. The reverse inhibition of the larger shoot by the smaller will obviously be less significant even than the inhibition between two equal shoots as in column 1; we arrive in this way at the required result that of two unequal axillary shoots the larger should be less, but the smaller more, unsymmetrical than two equal shoots would be. In the last column of Table 15 this principle is confirmed in experiments where leaf and bud have been destroyed together, so that the surviving axillary shoot, having no rival to exert any inhibition upon it, shows only the residual asymmetry which is impressed upon its by its own axillant leaf.

From the last line of the table it is apparent that a cutting from a young shoot can produce the normal set of symmetry responses only if the development of its root system is allowed to proceed unchecked, whereas in 'old shoot' cuttings the dependence of the correlation mechanism on root establishment is much less, though still noticeable. The importance of root growth for correlative effects in the shoot was further demonstrated in a series of experiments in which the isophyllous main axis of a young plant was deprived of all its mature leaves except one. The next leaf-pair to unfold in the plane of the surviving leaf becomes asymmetrical as a result of this operation, and the asymmetry can be expressed as a ratio of lengths in the usual way. For well-rooted plants the leaf above the surviving old one was 0·62, but for root-pruned stocks 1·12, times the length of its partner. In this form of experiment, therefore, the pruning of the roots actually reverses the correlative behaviour of the leaves. Further examination reveals a complicated physiological situation. Coumarin and IAA were applied to leaves in

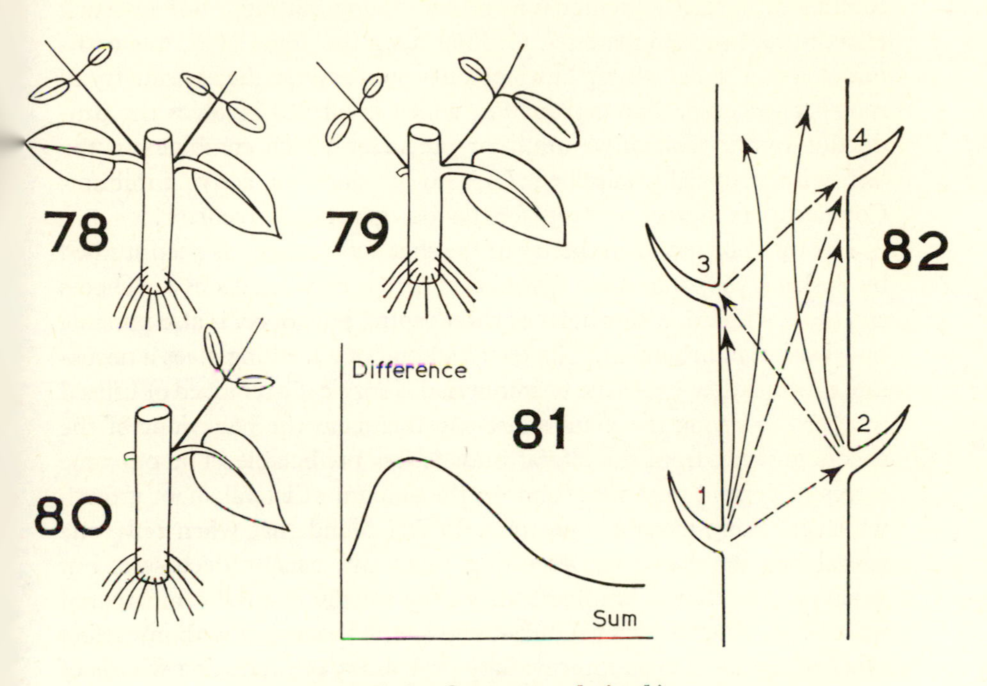

FIGURES 78–82. *Symmetry relationships.*

78–80. Rooted cuttings of *Coleus* subjected to surgical treatment (von Guttenberg & Müller, 1957). Each cutting has one node of the original stem. **78.** Intact cutting produces two axillary branches, on each of which the second leaf-pair shows moderate asymmetry (see Table 15). **79.** One original leaf removed, one axillary shoot with increased, the other with diminished, asymmetry. **80.** One original leaf and its bud removed, single axillary shoot with minimal asymmetry. **81.** Relationship between sum and difference (reckoned always positive) of prickle-counts from the two edges of a leaf in *Ilex* (simplified from Dormer & Hucker, 1957). Greatest asymmetry is found in only moderately prickly leaves (see text, p. 149). **82.** Scheme of communication in the shoots of distichous Leguminosae (Dormer, 1954; see also text, p. 152). Nodes are numbered, and each may be classed as R or L according to the position of its axillary buds. Each lower node, once its own condition has been settled, sends out impulses along the arrowed pathways, and these impulses characteristically change the probability of an R or L configuration at each upper node. As one possibility, let the continuous arrows selectively favour conveyance of R impulses, broken arrows that of L impulses. Any one node may be both a source, and a destination, for arrows of both types. It is therefore impossible for the filtering mechanism to be in the nodes; it must be in the tissues between them. Interchange of the two types of arrow (observed in some plants) leaves the force of this argument unimpaired.

concentrations of 10^{-8}, and it was found that IAA would produce a reversal of normal asymmetry in rooted young cuttings, but no such effect in root-pruned material. Coumarin, on the other hand, was without effect on rooted young cuttings but could reverse the asymmetry of root-pruned ones. The explanation which is offered involves the production by the roots of an inhibitor-precursor which combines in the leaf with artificially supplied IAA to produce an active inhibitor. Coumarin can in large part replace the root-generated precursor.

The variation in the symmetry of the shoot of *Catalpa* has been studied by Fosket (1968). The main distinction here is between decussate shoots and those with leaves in whorls of three; spiral phyllotaxy is also possible but relatively uncommon. The terminal bud of a seedling gives a decussate continuation, but if the terminal bud is surgically removed or caused to abort by giving the plant short-day treatment the behaviour of the shoots growing from the lateral buds is less predictable. The outcome seems to depend to some extent on the amount of lateral shoot growth which is taking place at one time. Fosket found that when only one lateral bud developed the resulting shoot was usually decussate, but when two developed simultaneously they usually had three-membered whorls. Seedlings which produced two lateral shoots but with imperfect synchronization had an intermediate probability of producing whorls of three. The situation is complicated by effects arising from the age of the parts, and will require further investigation.

Another study in which the initiation of a phyllotactic system appears to be influenced by a number of factors is the work of Rijven (1968) on various species of Leguminosae. The curvature of the embryo in the seed means that there are two possible positions (distinguished by Rijven as 'inside' and 'outside') for the first plumular leaf. It appears that different species and races produce varying proportions of individuals with 'inside' position. There is thus some evidence of genetical control, and claims were also made for differences brought about by changes in temperature and by supplying the plants with excess of calcium nitrate and magnesium chloride. In such a matter there is an obvious possibility that the condition of one embryo might be related to that of its neighbour in the pod. Rijven supposed he had found a negative correlation between adjoining seeds of *Trigonella*, but had really too few observations upon this point to yield any conclusive result. Examination (Rijven, 1969) of the position of the third plumular leaf, which may be either right or left of the first, reveals that the two possibilities, though perhaps equally frequent in large populations of seedlings, are not necessarily independent of other unilateral factors operating in the individual. There is evidence for a gravitational effect, with the third leaf appearing on the upper half in most though not all of the cases where an embryo lies on its side. Early surgical removal of one cotyledon tends to favour the appearance of the third leaf on whichever side is selected for the operation. This observation naturally suggests that natural inequalities between the weights of the cotyledons might be reflected in a tendency for the third leaf to occur preferentially on the same side as the smaller cotyledon. Rijven denies this, but on a number of dissections which is clearly too small to establish a negative conclusion of any real value.

The work so far presented in this chapter, though often leading to rough quantitative results, hardly progresses beyond the elementary stage at which methods of measurement are being introduced and given their preliminary trials. In many examples the number of observations has been insufficient to support any detailed analysis of the mathematical relationships involved, and the authors concerned seem generally not to have regarded the attainment of high levels of numerical accuracy as particularly desirable or necessary. Some other investigations, however, have been characterized by a greater abundance of observational data, and have opened up new prospects of interpretation.

A study by Dormer & Hucker (1957) of the marginal prickles on the leaf of *Ilex* revealed a complicated situation of which there are perhaps two aspects of general morphological significance. The observations were sufficient to permit a systematic examination of the relationship between the total spininess of the leaf on the one hand, and the inequality between the spine-counts of the two edges of the leaf on the other. The total number of lateral prickles on a leaf obviously sets a limit to the degree of asymmetry which that leaf could possibly display; a leaf with two lateral prickles can carry both on the same edge, giving an unbalance of two prickles, but lacks the equipment to generate any greater inequality. Development of this argument leads to the realization that leaves with odd and even totals of prickles must be considered separately, and much more importantly to an appreciation that conditions are here favourable to an assessment of the leaf's resistance to asymmetry. If there were no such resistance, then the mean difference between the prickle-counts of the two edges would presumably increase with the total number of prickles present, not necessarily in constant proportion but at least without any absolute ceiling of asymmetry ever becoming apparent. Another possible state of affairs would be the existence in the leaf of an equalizing mechanism which would permit the increase of asymmetry up to a fixed level of tolerance but no further. In the event, the situation in the holly leaf was found to be as in Fig. 81. The permitted amount of asymmetry increases with total prickle number up to a certain point, but thereafter actually declines. It seems inescapable that we should introduce in one form or another the concept that a given leaf-edge may become 'full' of prickles, and that the approach to this condition generates an increasing pressure for any additional prickle to be directed to the less crowded of the two edges. The same investigation led to a rather close consideration of the effect of the phyllotactic spiral upon the symmetry of the individual leaf. In principal one might reasonably anticipate that any such effect would be reversible with the phyllotaxy. If a shoot with right-hand phyllotaxy were held in front of a mirror, then in the reflected image both the phyllotactic spiral and any associated lopsidedness of particular leaves would be changed to left-handed equivalents. This is however a purely geometrical transposition; the question whether the mirror-image of a right-hand shoot is structurally interchangeable with a real left-hand shoot is an entirely separate matter, and one which clearly has to be submitted to the test of observation. The weight of evidence is against the existence of such interchangeability in holly. The two directions of phyllotaxy appear to be associated with different degrees of standardization of prickle numbers. The variances or standard deviations of the prickle-counts are non-stereo constants, the characteristic values of which are incapable of being interchanged by optical reflection. That properties which are not themselves of a right- or left-handed nature can be coupled in this way with the contrasting forms of unsymmetrical geometry is not particularly surprising, but recorded examples of such phenomena are rare, and some effort to increase their number would perhaps be appropriate.

Owing to the regularity of phyllotactic patterns, symmetry considerations are liable to arise where any effect is distributed in a periodic or repetitive manner along the length of a shoot. Brett & Dormer (1960) examined the fluctuations from node to node in the number of serrations on the leaf of *Spiraea*. The general situation is not very different from that shown in Fig. 52 (p. 96), with a large-scale seasonal rise and fall in serration numbers, upon which smaller and less regular variations are superimposed to give the graph a jagged outline. The oscillations in *Spiraea* did not however show any progressive diminution in amplitude, nor did they in other respects conform to any known pattern such as that arising from subdecussate phyllotaxy. It therefore became necessary to determine whether the variations were completely random, or whether the appearance of a leaf distinguished from its neighbours by a specially high or low count of serrations would tend to be followed at a characteristic interval by another leaf with the same peculiarity. Simple arithmetical devices, not in themselves of any general interest, permitted tests to be made for the existence of periodic tendencies, and the results were positive. In terms of a Fibonacci system it appears that any peculiarity of a particular leaf, whether it be specially rich or specially poor in marginal serrations, is liable to be propagated along a parastichy. The suggestion must be offered quite generally that when we mark out the parallel parastichies of any given set, the dimensions of any constituent leaf will not accurately conform to average values, but will be subject to small adjustments which will be persistently positive in some parastichies and persistently negative in others.

A periodic phenomenon which is very different in its structural manifestations, but which probably exemplifies the same underlying principle operates upon the distribution of flowers along the rhizomes of Nymphaeaceae (Dormer & Cutter, 1959). The development of a flower being an essentially qualitative change, the whole study must be conceived in terms of probability: the problem is everywhere one of estimating the probability that a particular node shall produce a flower. By counting along samples of stem the numbers of flowering nodes and of purely vegetative ones, we immediately obtain a general or average probability of flowering. As a first approximation this may be considered constant for a species, and it is the best estimate that can be given if the prospects of flower-formation have to be predicted in the absence of other information. The whole problem is transformed, however, as soon as the occurrence of a flower at one particular node has become an established fact, because an existing flower starts a complex oscillatory disturbance of the probabilities of flowering at the nodes which follow. Attach a zero to any flowering node as the starting-point for the system of numbering, and nodes 1, 5, and 8 will be found to have reduced prospects of producing a flower, while the probability of flowering at node 2 is in some species more than twice the general average for the rhizome. Each flower is in fact followed by a system of probability waves, to the action of which the mechanism of floral induction in other nodes is responsive. The wave originating from any flower can be followed without difficulty through a good many internodes, and through several alternations of crest and trough. The likelihood that such a wave may be propagated largely by resonance, with a flower A inducing another flower B in such a position that the wave emanating from B reinforces that emanating from A, is too obvious to be overlooked, and greatly complicates the problem of theoretical interpretation.

The advanced study of shoot symmetry, perhaps more than any other branch of plant morphology, appears likely to raise us to levels of mathematical abstraction which may fairly challenge comparison with those attained in more fundamental sciences. Total ignorance of the physical nature of the process of floral induction in *Nymphaea* need not inhibit any future investigator who wishes to examine the probability functions more closely; the probability of flowering possesses the essential attributes of wavelength, velocity, and amplitude, and is obviously a vehicle of energy transfer. In such conditions the familiar framework of wave-mechanics is not merely applicable but inescapable.

The question of communication, of the ability of one organ to influence the form of another situated at a distance from it, is one which constantly recurs in connection with investigations of symmetry. In favourable circumstances it is possible to effect a partial analysis of the communication system, as in a study (Dormer, 1954) of the shoots of various Leguminosae. The asymmetrical configuration utilized in this work was the arrangement of multiple axillary buds, which permits the nodes to be recognized as R or L according to an agreed convention. The two types are equally frequent but not randomly assorted: quite generally the probabilities as to the state of any upper node (to be called the receiver) are biased in a measurable way by the actual condition of any lower node (to be called the transmitter). Stripped of a good many complications which are inessential to the present discussion, what can be observed is a structural correlation which is intelligible only as a demonstration that an unsymmetrical impulse has passed from the transmitter to the receiver. The bias of the observed probabilities at whatever node may be selected as the receiver for a particular measurement enables us to assess the efficiency of the communication. If, for example, in a population of seedlings, the arrangement of the buds at the second plumular node were to be repeated at the fourth plumular node in 80% of specimens, but reversed at the third plumular node in only 65% of specimens, then we could argue that transmission from node 2 was twice as effective at node 4 as at node 3, the departures from 50% being 30%

and 15% respectively. The question of communication efficiency is evidently independent of the direction of receiver response: that node 3 tends to be R when node 2 is L, whereas 4 tends to be of the same sign as 2, is irrelevant to the consideration of the magnitude of the bias imposed on the receiver.

These ideas make it a simple matter to prepare charts showing the efficiencies with which the asymmetrical impulses are conveyed along the various routes from one node to another. When this is done for R and L receivers separately a curious conclusion emerges, in that the tissues of the plant appear to exert a filtering effect upon the messages passing through them. Along some of the pathways a transmitter which is R imposes a greater bias on the probable response of the receiver than does a transmitter which is L. On other routes the filtering is the other way round. The opportunity to make this analysis arises fortuitously from the fact, which could not have been predicted beforehand, that opposite types of filtration exist at the same time, in the same stem, and in respect of the same transmitters and receivers. The characteristic situation is shown in Fig. 82, where a particular direction of filtration cannot be attributed to R or L transmitters or receivers, but only to the pathways between them (see caption to figure).

When we survey the various systems of relationships which have been separately shown to operate in shoots, and when we reflect that two or more of them must very commonly be combined in the same stem, the resulting impression is one of very considerable complexity. The problems, though difficult, are not impregnable to experimental investigation or to mathematical analysis based on sufficiently copious observational data, and the subject appears to be moving out of the purely descriptive phase of its history. The search is no longer one for curious examples of lopsidedness, but rather for situations favourable to the application of intensive research techniques.

Chapter Six

The Vascular System

The number of vascular bundles in a transverse section of a stem is often roughly constant in successive internodes. As some bundles are withdrawn into lateral organs at every node, and as the known physiological functions of the vascular tissues demand a certain measure of continuity, we are to expect that the primary vascular system of the stem will constitute some kind of branching network. The study of vascular organization from this point of view has been pursued for more than a century but has never attracted large numbers of workers, perhaps because of the manipulative difficulties involved. The tedious routine of reconstructing vascular connections from serial sections can only to a limited extent be replaced by less laborious dissection methods. Simple vascular systems make appreciable demands upon the observer's patience; really complex ones can present tasks which exceed the normal endurance of an individual. Published accounts are therefore often incomplete, and many of their authors have had no opportunity to gain personal experience of any wide range of specimens. A common type of contribution is the description of a single species prepared by a worker for whom it is to be the only venture in this field. No appraisal of the literature can be satisfactory which does not consciously make allowances for the narrowness of the circle of vision within which many opinions have been formed.

In principle there are two main ways in which the examination of a vascular system can be approached. Investigation of a mature shoot must reveal the system as a network of communications, with some vascular bundles continuing from one internode through the intervening node into the base of the next internode while other bundles near them are diverted into the leaf or the axillary branch. Constituent bundles of the network, like the roads of a country, will vary in size and importance, and will at certain points branch, unite, or be linked by cross-connections. If there is to be any effective comparison between different types of construction some attempt must be made to recognize homologous parts, but the attachment of designations to strands which are physically connected must always involve a proportion of quite arbitrary or conventional decisions. The nature of the problem will be illustrated by the roads of any populous country which has given reference numbers to its highways: where two routes unite, it must be decided which (if either) of the available designations is to be allotted to the common trunk, and where a route forks it will be necessary to consider which (if either) of the 'new' roads is to be regarded as a direct continuation of the 'old' one. These matters cannot generally be resolved by any principle of geometry; history, convenience, or local custom may afford a useful but necessarily imperfect guide. As the whole of comparative morphology is in any case based upon the consistent application of nomenclatural conventions it is difficult to see why there should be any insurmountable objection to treating vascular arrangements in the same way.

Those workers who have concerned themselves with the development of the vascular system in the immature part of the shoot have however very generally adopted a different point of view. In connection with observations upon leaf primordia it is commonly found that some selected criterion of vascular differentiation, such as the first sign of xylem lignification, or even the first recognizable delimitation of procambium, is first seen in or near the leaf-base, and only later progresses downwards into the internodal part of the vascular bundle. There have always been anatomists for whom this basipetal sequence of development has represented the only acceptable foundation for the description of the vascular plan. Rigid insistence upon this idea leads directly to a denial that the stem in most angiosperms can possess any primary vascularization of its own at all. The bundles traversing the internode do not in any sense belong to it, but have grown into it from the leaves rather as roots grow into the soil. At times there has appeared the different but closely related doctrine that the stem, properly speaking, does not exist as an independent morphological category, but represents merely a sympodial union of leaf bases. While the essential facts concerning basipetal vascular differentiation do not appear to be seriously in dispute, there is no reason to suppose that the mode of development has any significant influence upon the activities of the vascular system at maturity, any more than passengers on a railway need to know the completion dates of the various bridges over which they pass. Nor can there be any security that the visual criteria by which the first 'appearance' of a vascular strand has been recognized will bear any fixed relationship to the physiological processes which have previously determined that the tissue shall become vascular. To observe the absolute beginning of differentiation might very possibly be the perfect method for studying the vascular arrangement; to observe what is thought to be the first visible change (which is all that can be done in practice) is not necessarily an adequate substitute.

The description of vascular systems in terms of basipetally-developed leaf traces has certain practical disadvantages. It often involves the application to a geometrical pattern which is essentially static of words which carry implications of motion. Reynolds (1942) for instance, in describing the vascular system of *Ricinus*, has leaf traces 'descending' the stem and being 'blocked by incoming vascular bundles from the lower leaves'. Such expressions have an emotional content which is entirely false. Even with the greatest respect for the principle of basipetal development, a leaf trace can only 'descend' in that abstract and figurative sense in which a road 'goes' to a town; the vascular bundles associated with the lower leaves are only 'incoming' by some such convention as that by which English railwaymen have persuaded themselves that trains always go 'up' to London but 'down' from it. Nor is it immediately clear how one stationary piece of tissue can 'block' another. Without necessarily wishing to eliminate all vivacity from scientific writing, one may question the wisdom of choosing such energetic language for tasks of routine morphological description. Written accounts which follow the 'descending trace' convention also very commonly fail to convey any clear impression to the mind of a reader who does not concurrently refer to a diagram; if correctly stated the description will embody all the instructions needed for the construction of a diagram, but that is not quite the point. If a system of description cannot convey its meaning without being subjected to a kind of geometrical decoding process, then its functional efficiency is not very high. By this test the concept of the descending trace, as a device for imparting information, leaves much to be desired.

Regardless of their relative merits, however, different techniques of description, if correctly executed, ought to result in factually equivalent records. Published accounts may differ profoundly in their underlying philosophies and may define such terms as 'leaf-trace' in incompatible ways, but it must always be possible for a person familiar with the conventions employed to translate the rival descriptions to some common ground and there check their agreement point by point. Indeed, without some willingness to make this kind of adjustment it is impossible to attain a general knowledge of the subject, the use of some common expressions such as 'leaf-trace' never having been effectively standardized. Unfortunately some workers have not clearly distinguished questions of fact, about which there should rarely be any dispute, from questions of descriptive usage, which must be in the last resort a matter of taste. An example of this confusion can be seen in the work of O'Neill (1961) on the shoot of *Lupinus*. He worked with the extreme tip of a shoot, found, as was to be expected in these circumstances, a predominantly basipetal sequence of development, and accordingly described the vascular system as a set of leaf-traces. His conception of the system is essentially that shown in Fig. 83 (see p. 158), where vascular bundles are formed progressively downwards from the numbered leaves into the stem. Each bundle, after running a free course through several internodes, joins another according to a pattern (related to the phyllotaxy) which is somewhat irregular but by no means completely random. From O'Neill's text we can extract two forms of statement: 'If followed upward, the traces of a given leaf appear as branches of traces of other leaves' and 'The traces in a sequence constitute an interconnected series that may be called a sympodium.' Clearly there is some difficulty here in finding a suitable name for any strand which has upward connections with more than one leaf. In some earlier publications (Dormer, 1945, 1946, 1954a) the expression 'stem bundle' had been proposed as a convenient and developmentally neutral term for any bundle of this type. O'Neill will have none of this. The diagrams showing stem bundles, he says, 'do not resemble the vascular systems of *Lupinus* and most other dicotyledons. These systems consist of strands that are related to leaves as leaf traces and their complexes. There are no indefinitely prolonged stem bundles.' The objection here is not however to the structure but to the name which had been suggested for it. Call the compound strand a stem bundle and its existence is denied; allow it to pass as a sympodium, or as a complex, or possibly even as a special kind of leaf-trace which can serve more than one leaf, and O'Neill is perfectly content to admit it as an essential part of his scheme of vascular organization. There have been other instances where an apparent conflict as to facts has in reality involved nothing more serious than a verbal misunderstanding, or (which may have been a significant factor in the O'Neill example also) a failure to allow for incompleteness of one diagram as compared with another. A drawing prepared from a very young stem is likely, for instance, to exaggerate the distance through which a trace remains

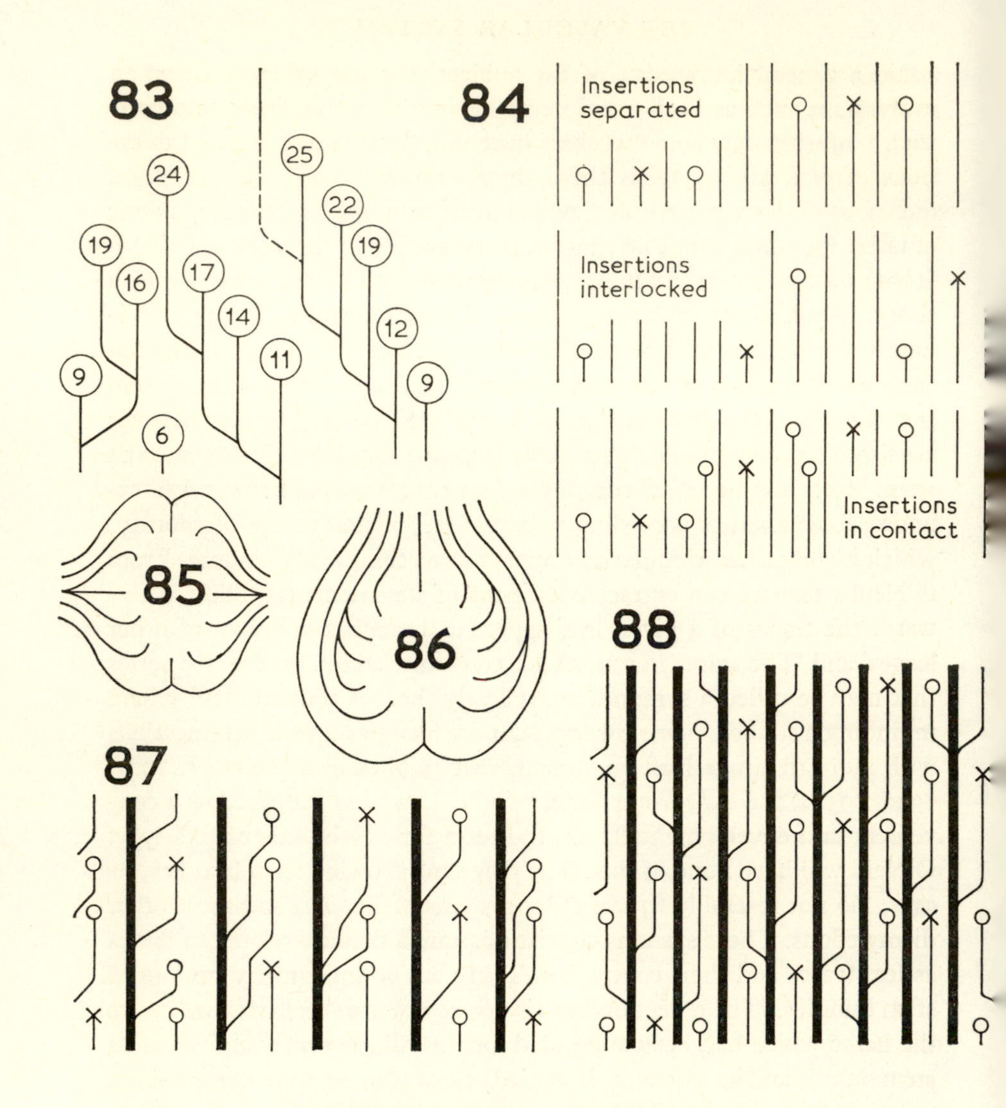

FIGURES 83–88. *Primary vascular construction.*

83. Transcription of part of a diagram by O'Neill (1961), each trace labelled with the number of its leaf, the broken line added to the original to represent a probable later development. In one terminology trace 25 runs down to a junction with trace 22; in another, the lower part of O'Neill's 'trace' 25 will be shown by the later addition to be a portion of stem bundle. **84.** Stem vascularization, cross denoting median leaf trace, circle a lateral trace. Three types of relationship between successive trilacunar leaf-insertions. 'Insertions in contact' covers two minor variants. **85.** Plan view of traces running out to paired leaves in Valerianaceae, etc., two traces shared between the leaves. **86.** Analogous situation for single leaf in Magnoliaceae. **87 & 88.** Two open vascular systems with

completely independent, because it will represent a stage previous to the appearance of some of the upper connections. In Fig. 83 the broken line represents a probable vascular connection which had not become recognizable at the time when O'Neill took his observations.

In the comparative study of vascular systems it is necessary to distinguish several possible arrangements of the traces of any one leaf. The survey by Sinnott (1914) showed that there were three principal patterns: the unilacunar node with a single leaf gap, the trilacunar node with three, and the multilacunar node with more than three. Although later work has revealed a greater measure of variability than Sinnott supposed to exist, it remains basically true that a particular nodal state is generally characteristic of large taxonomic groupings such as families; with certain exceptions the Umbelliferae are multilacunar, the Ericaceae unilacunar, the Leguminosae trilacunar, and so on. In the majority of cases there is only one leaf-trace in each gap, but the occurrence of multiple traces in a common gap is a feature of certain families such as Cruciferae (Ezelarab & Dormer, 1966). It has long been apparent that a broad correlation exists between the vascular condition of the node and the external form of the leaf-base. Typically a unilacunar node implies an exstipulate leaf, a trilacunar one a leaf with distinct stipules, and a multilacunar node a more or less sheathing base. The common mode of origin for the veins of a stipule is by branching from a lateral leaf-trace as it runs through the cortex or the base of the petiole. Examples exist (Dormer, 1944) in which a leaf has a lateral trace and gap on one side only, and only the corresponding stipule, while in *Acacia verticillata* there are two kinds of leaf, one with a trilacunar node and stipules, the other with a unilacunar node and no stipules. In many species, more especially those with a large number of traces, the vascular condition of the node is subject to considerable heteroblastic modification as the growth of the shoot proceeds. Where a species is characterized by a highly multilacunar state it is not to be expected that the full complexity of which the plant is capable can be attained in the first few nodes of the plumule, or indeed of any lateral shoot. One must also be prepared for progressive simplification of pattern on approaching an inflorescence.

---

5 and 8 stem bundles respectively, both with regular spiral phyllotaxy, based on fractions 2/5 and 3/8. Stem bundles identified by thicker line, but not necessarily larger or straighter than traces in actual specimens. These systems are regular (but not identical) in direction of trace-origin, but realistically variable in lengths of free course of traces. Items on the edge of each diagram are shown twice.

While heteroblastic changes in vascularization may be of interest in themselves or as subjects for physiological experiment, attempts to utilize the facts of vascular construction for comparative purposes must in the main be based upon the most elaborate nodal arrangements which a species can produce. It is there that the genetical influences will be most fully expressed.

The way in which foliar traces are derived from the vascular system must depend not only upon their number but also upon their circumferential spacing, and this in turn will obviously have some relationships to the phyllotaxy. There is no generally accepted style of description for dealing with this aspect of the problem. Other things being equal, the fraction of the circumference which is occupied by the vascular attachments of one leaf will tend to vary directly as the number of gaps. Many unilacunar nodes have only a single trace, but a multilacunar node may have traces spaced out round the whole circumference, indeed in some monocotyledons the edges of the sheath are overlapped so that the leaf-insertion occupies more than 360°. No universal rule can be established, however, and those unilacunar nodes in which the gap contains several traces may have a considerable width of insertion. In relation mainly to trilacunar nodes, and more especially to those which occur in a spiral phyllotactic sequence, it is possible to apply the nomenclature (Dormer, 1945) shown in Fig. 84. This requires the tracing of bundles through the internode, but has the advantage of stating morphological distinctions which do not depend on estimates of angular divergence and are unchanged by accidental deviations in the course of the bundles. Some interesting cases exist in which a bundle which emerges from the vascular system of the stem as a single strand appears, from its behaviour, not to be of the nature of a simple leaf-trace, but rather to represent a fusion of two. In Valerianaceae, Dipsacaceae, and Rhizophoraceae, where there is decussate phyllotaxy, there are traces which are shared between the leaves of the pair (Fig. 85). Some Magnoliaceae show an equivalent situation with only a single leaf (Fig. 86).

So far as the more detailed analysis of stem vascularization is concerned, there appears to be a growing recognition that the exact level in the stem at which a particular bifurcation or recombination of bundles takes place is not, in the general case, an important morphological constant. The trace of one leaf, if followed downwards, may run through several internodes before joining any other strand, while the corresponding trace of the next leaf has a much shorter free course. Only in a minority of systems, where exceptional geometrical regularity prevails, are such things at all precisely regulated. It is this variation which renders so unsuitable for comparative purposes the type of description which is inspired by the downward tracking of individual strands. In attempting to reach a broader view we may draw a distinction (Dormer, 1945) between 'open' vascular systems in which the bundles in their upward course branch but only exceptionally rejoin, and 'closed' systems in which there is a general equilibrium between branching and anastomosis, so that the system assumes the properties of a network. The difference certainly is not absolute; systems exist in which anastomoses and cross-connections occur with intermediate frequencies. There are many observations, mostly recorded incidentally in the progress of other work, suggesting that the cross-connections which complete the network in closed systems are often later in their differentiation than the other bundles, and certainly there are many examples in which the bundles responsible for cross-linkage are of conspicuously small calibre. There is, in short, a mass of detail which might perhaps fall into a coherent scheme if we adopted the hypothesis that cross-connections had been introduced progressively into systems which were perfectly open at the commencement of the evolutionary process.

Vascular systems of the open type, which in some important reviews (Balfour & Philipson, 1962; Philipson & Balfour, 1963) has been called the 'sympodial' type, are very widespread. In those which have been followed for a sufficient distance for all the bundles to be accounted for, it has generally turned out that all the traces spring from a definite, and often rather small, number of stem bundles or sympodia, all of which are indefinitely continued upwards. It is a convenient though by no means universal practice to distinguish these fundamental strands in diagrams by a bolder line (Figs. 87, 88). They are not, however, necessarily larger than trace bundles, and are not always distinguished from traces in any way which can be infallibly recognized in a section. It follows therefore that the upper part of such a diagram can only be completed with the aid of observations taken at still higher levels; where a bundle forks it may be impossible to decide which portion is trace and which is stem bundle until both have been followed to their destinations. The surface of the vascular cylinder effectively consists of a set of panels, each bounded by two stem bundles; in some instances this pattern is related to the surface sculpturing of the stem. The panels are treated in our diagrams as running vertically, but it is likely that all the bundles pursue a slightly helical course; in other words there may be an angular discrepancy between the actual phyllotaxy and the approximate fraction (e.g. 2/5 in Fig. 87) which would be deduced from the arrangement of the bundles. The very gentle spiral winding of the vascular system which is required to correct this discrepancy can for many purposes be neglected. The systems shown in Figs. 87 & 88 are of a common type in which each panel usually contains only one trace bundle at any given level. Consequently the total count of bundles at the middle of an internode approximates to twice the number of stem bundles. Any distortion which results in points of bifurcation being moved upwards will result in a reduction in the average bundle-count, while if the points of branching are moved downwards the total number of bundles in a transverse section must tend to increase, which, in an evolutionary context, would be likely to lead to some enlargement of the pith. The appearance of the internodal section would be very similarly affected if the number of stem bundles were to be increased. Girolami (1953) has reported in *Linum* a system with twenty-one stem bundles and we have no reason to deny the possibility of even larger numbers. These considerations are of some importance because there are many dicotyledonous stems with bundle-counts of fifty or more. We are not yet in a position to assess the probabilities regarding the evolutionary history of this many-bundled condition.

The open systems, both as a group and individually, are variable not only in the level at which vascular bifurcation takes place, but also in respect of the direction in which a trace springs from its parent stem bundle. Figs. 87 & 88 have been constructed to illustrate some possible governing principles, as for example that traces shall take their departure in the ascending sense of the phyllotactic spiral, or that a lateral trace shall be given off in the direction towards its associated median trace. Such generalizations are not, however, consistent from one species to another, and the behaviour of some systems is so erratic that any detailed enquiry would necessarily assume a statistical aspect. Some published illustrations almost certainly show a degree of geometrical regularity which did not exist in the specimen. Some of the earlier authors, doubtless in good faith and without understanding the uncertainties involved, appear to have resorted to the diagrammatic repetition of a pattern of construction which had only been directly observed in rather too short a piece of stem. In more recent times those studies which have been closely related to apical differentiation, and which have consequently employed material in which only negligible internodal elongation had occurred, have necessarily been conducted in conditions unfavourable to the estimation of longitudinal distances. Where the length of every internode is virtually zero, differences in the length of free course of traces will escape attention.

Despite the variability of open systems in detail, some of them display marked stability in the number of their stem bundles and in the geometrical relationship between the insertions of successive leaves. It is possible indeed to recognize a common pattern (the 'acacian' system of Dormer (1945), the definition of which is independent of the detailed behaviour of particular strands. Essentially this is the spiral trilacunar type with stem bundles in Fibonacci numbers (Figs. 87 & 88) and with approximately one trace in each panel at each level. Such systems are found in diverse groups of dicotyledons, and the type has very possibly been the starting-point for various evolutionary sequences; it does not follow that it is absolutely the primitive form for the angiosperms as a whole. A variant has been described (Dormer, 1954a) characterized by the presence of an intermediate category of bundle, the 'trace-complex'. A trace-complex behaves for a time like a stem bundle, giving off traces as branches, but its persistence is limited, and ultimately it passes outwards as a trace. A trace-complex is perhaps not a very rare phenomenon and its existence is probably related to the phyllotactic discrepancy mentioned on p. 162. Although there seems to have been no investigation primarily directed to the elucidation of juvenile stages, there have been enough casual observations to show that a shoot in its early growth, whether a plumule or a lateral branch, very often has a deficiency of stem bundles which has to be adjusted (by bifurcation of one or more of the stem bundles available) before a fully adult vascular pattern can be established.

While many closed vascular systems undoubtedly display an inconstancy of detail which is entirely comparable with that of the open systems, the closed systems which have been most completely analysed have often revealed a remarkable fixity of organization. Thus in a number of small herbaceous Leguminosae Dormer (1946) found the patterns shown in Figs. 89 & 90 (see p. 164), as well as related forms differing from these in the number of bundles in the internode. In shoots of this type the internodal bundle-count remains constant for long distances; in some cases it is, so far as is known, absolutely constant for the species, whereas other examples show constancy of structure within a given

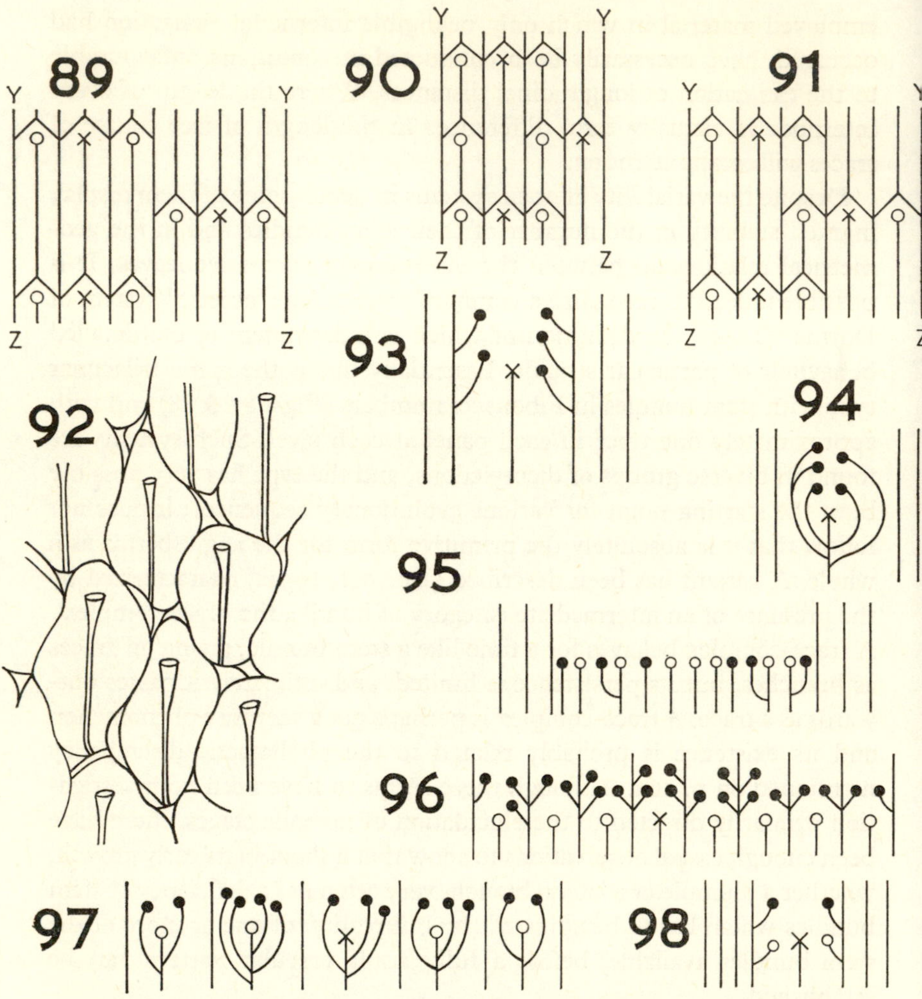

FIGURES 89–98. *Vascular systems viewed through the surface of the stem.* Cross denotes a median leaf trace, circle a lateral leaf trace, black dot a branch trace. **89–91.** Closed systems of various Leguminosae, in each of which the bundle YZ is shown twice. **91** is the dorsiventral system in the prostrate stem of *Scorpiurus*, where YZ lies against the soil. **92.** Major vascular strands in stem-joint of *Opuntia*; each of the larger meshes here contains a single trace serving one areole, and a much finer subordinate reticulation, not shown here, fills in the spaces between the principal strands. **93.** Normal origin of axillary branch traces. **94.** Axillary supply in *Berberis*. **95.** Axillary bud traces in *Lepidium*, intermingled with and in part derived from leaf traces which share a single wide gap and among which it is impossible to distinguish a median. **96.** Axillary supply in *Petasites*. **97.** Axillary supply in *Thalictrum*. **98.** Leaf traces arising in the manner of normal branch traces, as in *Impatiens* and some others.

stem, but differences (presumably reflecting inequalities of vigour and apical size) between stems of the same plant. A main axis with twelve bundles in every internode may, in these variable species, have lateral branches with ten or eight, and so on. These Leguminosae were all distichous and while most of them had approximately symmetrical vascular systems the markedly dorsiventral shoot of *Scorpiurus* displayed the vascular asymmetry of Fig. 91, where lateral leaf-traces on the gravitationally upper side of the stem have only half the free course of those on the lower side. Smith (1928), in a survey of 138 species of *Clematis*, showed that similar numerical regularities may exist in plants with decussate phyllotaxy. It would be unsafe, however, to forecast the position in closed systems of greater complexity; we know very little as yet about closed systems with more than three traces to a leaf or more than twelve bundles in the internodal section. Although it is inherent in the nature of a closed vascular system that the bundles should form a network in the stem, this state of affairs is not usually conspicuous in the examination of specimens; bundles in an internode generally run straight and parallel and cross-connections are often quite sharply restricted to the vicinity of the node. The existence of closed meshes, however, affords an opportunity to develop a more obviously reticulate venation in any case where there may be any pronounced dilatation of the stem. *Opuntia* is an example: if one of the flattened stem-joints be stripped of its cortex (an operation which is facilitated by preliminary soaking in hydrochloric acid) the vascular system which is revealed possesses a primary reticulation of the form shown in Fig. 92. Each 'cell' of this network encloses one areole, served by the trace which is indicated, and is filled also with a much finer subordinate reticulation, not shown in the diagram.

The vascular supply to an axillary branch is arranged upon a different basis from that to a leaf. Whereas in a majority of dicotyledons the leaf-traces have an extended free course in the stem, and can be recognized a long way below the node at which they emerge, an internodal section in a typical dicotyledon contains no strand which can reasonably be called a branch trace. There may, of course, be individual vessels which are destined to pass at a higher level into a lateral shoot, but these are completely assimilated into the other bundles of the system. They do not, at the internodal level, run with an independent group of 'branch-destined' protoxylem. This is why, in comparing different types of vascular system, we have so far been able to proceed without considering the branch supply. In the great majority of dicotyledons the entire primary vascularization of the branch (or branches, if there is more than one) springs from the bundles adjacent to the median gap of the axillant leaf, as in Fig. 93. The number of branch traces is variable, even within an individual. Often there are obviously more than two, but there are not always enough for the axillary shoot to start life with the full complement of bundles which it will possess in adult form. Because of these arrangements the vascular connections of a lateral branch with its parent stem will commonly extend over a smaller part of the circumference than do those of the axillant leaf.

Other modes of origin for branch traces have been discovered in a minority of dicotyledons. It is convenient to designate these vascular conditions by the names of genera in which they have been shown to occur, subject, however, to the proviso that even within a single node there may sometimes be a mixture of different types of vascular connection. The principal variations may be arranged as follows:

(*a*) *Alternative modes of branch-trace origin within the median gap of the axillant leaf.*

1. *Berberis*-condition (Dormer, 1954a, and Fig. 94). Branch-traces springing from the median leaf-trace.

2. *Lepidium*-condition (Ezelarab & Dormer, 1963, 1966, and Fig. 95). Branch-traces intermingled with, and springing from, multiple leaf-traces within the median foliar gap. In this type it is not always possible to designate any particular leaf-trace as the median one.

(*b*) *Constructions involving the origin of branch-traces from gaps other than the median one.*

3. *Petasites*-condition (Dormer, 1950, and Fig. 96). Branch-traces arising from at least some of the lateral leaf-gaps, in a manner similar to the normal origin of traces in a median gap.

4. *Thalictrum*-condition (Ezelarab & Dormer, 1963, and Fig. 97). Similar in principle to the *Berberis*-condition, but with branch-traces arising from lateral leaf-traces as well as from the median one.

The derivation of branch-traces from more than one foliar gap often implies a conspicuous development of strands which, in order to enter the base of the branch, will run almost horizontally round the node. Dormer (1950) has figured a case in which the branch stele has a massive and complex basal portion which embraces the vascular system of the parent stem, and which in transverse sections will assume the shape of a crescent.

The nature of the axillary supply, like other aspects of vascular organization, is of some taxonomic significance. The *Petasites*-condition, perhaps the most widespread of the specialized types, is thought to be prevalent in Umbelliferae, Araliaceae, and Plumbaginaceae, occurs less consistently in particular members of a few other dicotyledonous families, but has not otherwise been encountered. Similarly the intermingling of leaf- and branch-traces seen in *Lepidium*, though common enough in Cruciferae and Papaveraceae, is foreign to the behaviour of most other dicotyledons. In making comparisons it is necessary to work from the most complex nodal structure which the species can normally produce. The taxonomic character will be the inherent ability to produce a *Petasites*-condition; the fact that this potential may not be realized in every node need not obscure the significance of the observation.

Some plants exhibit features of nodal construction which raise the possibility that the destination of a strand has undergone some evolutionary change, as for example by the diversion into a leaf of a vascular bundle which originally served its axillary bud. Among the Tubiflorae, the dicotyledonous order which includes such families as Labiatae and Scrophulariaceae, a leaf may be served by three bundles in the manner of Fig. 98. The median trace has here a considerable free course in the stem but the laterals arise in the subnodal region, which may generate a suspicion, though it clearly does not amount to proof, that they are displaced branch-traces. Similar situations can be found elsewhere, as for example in *Impatiens*. The Verbenaceae and Labiatae present a special case in that the central bundle of the trio (possibly the only original foliar trace) has here tended to become very small. It is the little strand which appears at the middle of each side in transverse sections of the approximately square internodes so characteristic of these families. In extreme cases it has vanished entirely, as in some Verbenaceae described by Shah (1968). Here the vascular supply to the axillary buds is drawn from the foliar strands in a manner for which conflicting explanations could readily be offered (Fig. 99, see p. 168). A diversion of traces from one bud into another has possibly occurred in the evolution of *Asarum* (Dormer, 1955). The shoot system here is sympodial, and the phyllotaxy distichous. Each season's growth consists of a short stem which has only two foliage leaves, though it also possesses a few scale leaves. The two foliage leaves of the year are separated by an exceptionally short

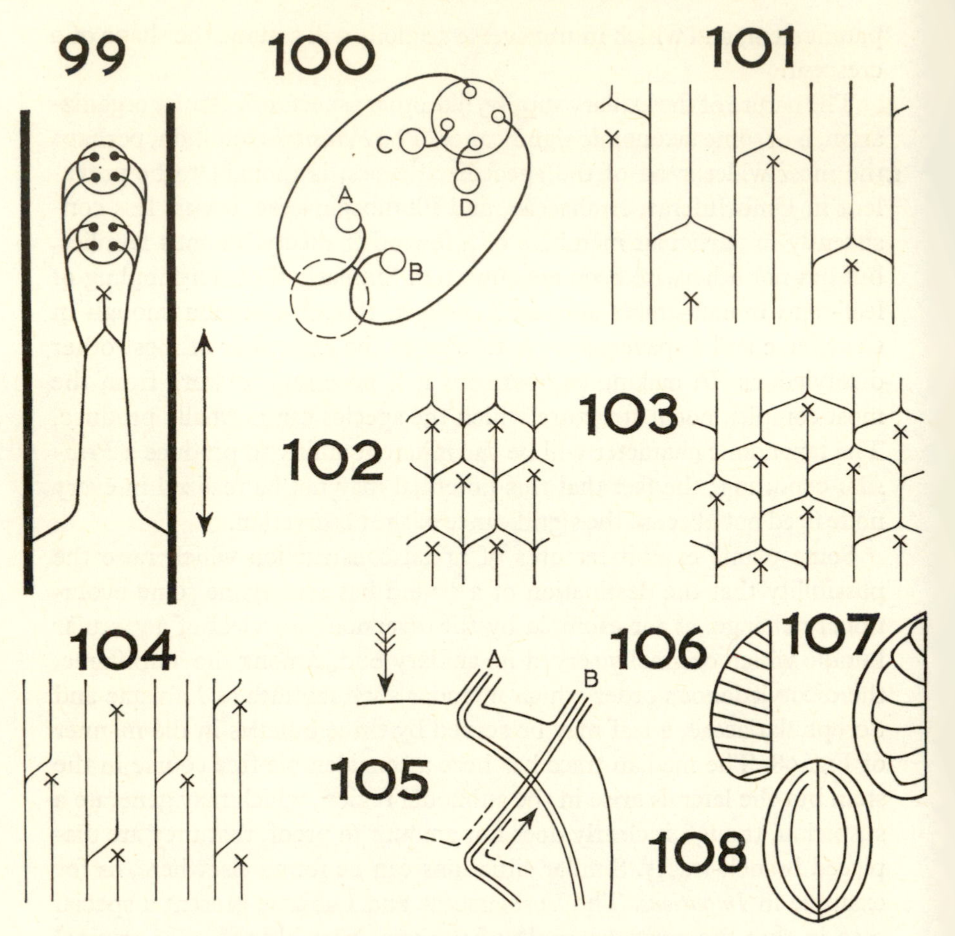

FIGURES 99–108. *Details of vascular construction.*

99. Vascular supply to axillary buds in *Clerodendrum* (Shah, 1968), the dimension indicated at the right being almost the length of an internode. **100.** Plan view of axillary traces associated with two imperfectly opposite foliage leaves in *Asarum* (Dormer, 1955). The broken outline is the expected position for a bud of the lower leaf, to be vascularized by branches from the stem-bundles A and B; in fact there is no such bud, and the traces continue to the far side of the stem and enter the bud of the upper leaf, along with the expected contributions from C and D. **101–104.** Vascular systems of Chenopodiaceae (Fahn & Arzee, 1959; Fahn & Broido, 1963). **101** in *Salsola soda*, **102** in *S. rosmarinus*, *Anabasis* and *Haloxylon*, **103** in *Suaeda palaestina*, **104** in *Arthrocnemum*, *Salicornia*, *Halocnemium*. **105.** Origin of characteristic bend of the leaf trace in thick-stemmed monocotyledons. Feathered arrow indicates apex of stem seen in longitudinal section. A is a young leaf primoridum, B an older leaf. An earlier position of B is shown by broken lines; B has moved in the direction of the small arrow. Vascular bundles shown by double lines. The bend arises just below the apical sur-

internode, making them sub-opposite. Their relationships as to axillary vascularization are shown in Fig. 100. The upper leaf of the pair has an unusually large axillary bud which carries the special responsibility of continuing the sympodium in the following season. The lower leaf has no axillary bud at all, but bundles arising in the places where bud-traces would be expected to appear run round to the other side of the stem and contribute to the supply of the enlarged bud of the other leaf. This arrangement is clearly exceptional (for the scale leaves have a perfectly normal pattern of bud supply) and may be provisionally regarded as a case of capture by one bud of traces which originally served another, the change being occasioned by an easily recognizable physiological need.

In most dicotyledons the form of the primary vascular system in the stem appears ill-adapted to the communication functions which it presumably has to serve. It will be sufficient to draw attention to three specific weaknesses of design: (*a*) although some part of the photosynthetic output of an older leaf may well be required to support growth at a higher level in the stem, material issuing from a leaf must generally pursue a downward course of some length before it can gain access to any strand which extends above the node at which the material was produced; (*b*) although a lateral branch may need to draw substantial supplies from its parent shoot it typically has direct connections with only two neighbouring vascular bundles in the latter, no matter how many bundles the main shoot may contain; (*c*) although many circumstances would seem to call for transference of materials round the stem in a tangential direction, an open vascular system makes no provision for such transference to take place within the stem at all (in some instances the movement might be accomplished by way of a leaf, taking substances out of the stem, through connections in the petiole, and so back to the stem on a different radius). All such limitations of vascular continuity must be expected to lose their effect when continuous layers of secondary layers of secondary xylem and phloem are established, but there is a prima facie case for supposing them to have some importance in young shoots or in those where secondary vascularization is delayed or diminished.

The known distribution of different types of vascular system tends to

---

face as a result of the rapid sub-apical enlargement of the pith, and comes to be buried deep in the stem because it does not participate in the outward movement of the leaf. **106–108.** Alternative arrangements of the principal lateral veins in an elliptical leaf lamina.

support the idea that considerations of tangential continuity have had some importance in regulating the evolution of shoots. Typical open systems are found only in plants where development of interfascicular xylem and phloem is copious and reasonably early. In monocotyledons, and in those dicotyledons in which the interfascicular tissue is absent, or very late in appearance, or of a purely parenchymatous nature, there is always some provision of vascular connections leading round the stem and permitting some measure of communication between the different vertical strands. This provision can be made in various ways, by the regular closing of leaf-gaps which characterizes the closed systems, by the development of girdling branch traces as in *Petasites*, or by special types of vascular anastomosis, examples of which are described later in this chapter. There is no particular reason to suppose that the need to maintain tangential continuity of the vascular system has ever imposed any serious restraint on the ecological adaptability of plants. Plenty of species which are thoroughly herbaceous in habit possess a vascular organization which exemplifies all the functional deficiencies of an open system, but continue to produce an adequately continuous cylinder of secondary tissue. It is also quite clear that closed systems, and some other types of more continuous primary construction, are by no means confined to plants in which high functional significance can be plausibly attached to them. A woody shoot with abundant secondary tissue presumably has no very urgent need of primary cross-connections, because other woody plants of similar habit survive without them, but it does not follow that such cross-connections will not be present. The permissible scope of generalization appears to be that cross-connections of various kinds appeared, were not necessarily of great functional importance at the outset, but created, when present with sufficient frequency and regularity, an opportunity to diminish the production of interfascicular tissue, without harmful consequences. The mere occurrence of such an opportunity would not guarantee its exploitation; some plants, though equipped with a primary vascularization which was potentially self-sufficient, would continue with normal secondary development. Others, however, were induced to start that progressive elimination of cambial activity which terminated in the monocotyledons and in such dicotyledons as *Peperomia*.

The principles of vascular morphology which have so far been introduced can be applied directly to the comparative study of examples within a family or larger group. One of the best surveys is the examination of Crassulaceae by Jensen (1968). Some of Jensen's diagrams are of unusual length, showing all the vascular connections through distances of more than twenty internodes. In this family there is considerable diversity, some species having open systems, others closed ones, while some are in an intermediate state with closure of some foliar gaps but not others. There are also species with trilacunar nodes, and unilacunar ones without lateral traces, as well as some in which, as in some Rhoeadales, lateral traces can share the gap of the median trace. The Crassulaceae are evidently comparable with other families in their capacity for producing individual variations upon a given basic type of vascular construction, and they also show marked changes in the vascular plan of a shoot during its successive stages of development. Thus for example in a plant of *Bryophyllum* (called *Kalanchoë* in the original) grown from one of the foliar buds, the system progresses within one of Jensen's diagrams from a juvenile to an adult condition. Except that each leaf has three traces throughout, the upper and lower parts of this stem are very different. The lower portion has decussate phyllotaxy with eight stem bundles (Jensen calls them sympodia), the upper part has spiral phyllotaxy with thirteen, which is one of the largest numbers reliably established in any stem. Whereas in the lower part it is common for a lateral trace to share the gap of the median, in the upper nodes this condition is excluded. There are other differences between the various levels, and the increase in the number of stem bundles is accomplished in an intermediate region in a manner which involves the closure of several median gaps in a system which is otherwise almost uniformly open. Even in the region which has settled to a stable form of spiral phyllotaxy with a constant number of stem bundles, however, there continues to be fluctuation in the length of free course of the various traces, and there are occasional irregularities in their direction of origin. It appears indeed that many Crassulaceae are sufficiently variable in structure to rule out any possibility of writing a readily intelligible account in terms of descending traces. Such variations, however, offer no real impediment to comparative study, and Jensen was able to bring vascular evidence in support of several taxonomic suggestions proposed on other grounds, among them a recognition that there are two distinct sections in the genus *Cotyledon*, while the separation of *Bryophyllum* from *Kalanchoë* is difficult to justify.

In a study of a range of conifers Namboodiri & Beck (1968) have reconstructed very regular open vascular systems for all the species with spiral phyllotaxy which they examined. This work is interesting because it included some bijugate specimens, with stem bundles in such numbers as 10 and 16 (which are Fibonacci numbers doubled) and one specimen from an accessory series, in which there were eighteen stem bundles or sympodia. This investigation, however, was evidently attended with certain technical difficulties arising from the relative lack of space between the bundles. Acceptance of the results involves acceptance of the principle that two 'sympodia' may come into actual tangential contact, and then separate again at a higher level, without having joined in any sense which would justify recognition as the closure of a gap. This contention is not necessarily unsound, but as no similar state of affairs has been admitted in the case of any angiosperm the situation must be viewed with some reserve until the histological aspects of the matter have been more fully examined. In this respect it is possibly a disadvantage that Namboodiri & Beck worked with apical buds where the vascular tissues were necessarily in an immature condition. The distinction between 'tangential contact' of two sympodia and the more intimate communication of a sympodium with its derivative traces was clearly regarded by the authors themselves as a delicate point, and any confusion would be very damaging to the perfect regularity which they attributed to these vascular systems. In dealing with conifers having verticillate phyllotaxy they found no difficulty in distinguishing between open and closed systems (both types occur), and the bundles in the verticillate forms are so much less crowded as to reduce very considerably the uncertainties of interpretation. Among the conifers with whorled leaves and open vascular systems the direction in which a trace branches off from its parent sympodium is often variable.

In most of the vascular systems so far discussed the strands in the stem are arranged in a tubular surface. That is to say, if the stem were to be slit open down one side and unrolled, in accordance with the convention adopted for many of the accompanying diagrams, then all the strands could be brought to lie in a single flat surface. A little plastic deformation would no doubt be required, but no strand need be ruptured and no strand need pass over or under any other strand upon the proposed test surface. In herbaceous stems of suitable texture this dissection can be performed in reality. A stem which presents a more complex appearance may nevertheless preserve the principle of tubular construction; in a number of cases the outgoing leaf-traces, without other abnormality of behaviour, make an outward turn into the cortex some distance below the node of the leaf they are to serve. An internodal transverse section may then show a distinctive anatomical construction with two rings of bundles, but the geometrical change which is involved is really of a most trivial character. Many stems, however, possess a vascular system which genuinely departs from the tubular form; their strands constitute a three-dimensional sponge and are interlaced so that no amount of plastic deformation will permit all the bundles to lie against a test surface. The difference is fundamental and is of a geometrical nature which admits of no compromise. Bundles in the more complex systems are crossed, and two bundles which cross cannot be brought into a single surface of any shape unless one of the bundles is severed. Challenged to reduce a given tangle into a flat layer of bundles, we would have a choice as to the position of our cuts, but would be unable to reduce their number below a certain minimum, this minimum number being a topological constant of the given problem. The vascular anatomy of monocotyledons and certain other plants is thus connected with the mathematical theory of string-games and knots. Even without formal mathematical treatment, it is possible to assess the relative geometrical complexity of different examples.

In several families of dicotyledons, including especially the Chenopodiaceae and Amarantaceae, there occurs a mode of secondary thickening resembling that of such monocotyledons as *Pandanus* and *Dracaena*, with a ring of meristem, external to the primary vascular system, which produces internally a mass of sclerotic conjunctive tissue in which complete collateral secondary bundles are embedded. Some of these plants show in addition a disturbance of the primary arrangement, some of the primary strands being in a 'medullary' position. The diagrams of vascular connections given by Wilson (1924) show that this is often a very trivial modification; the stem bundles lie nearer to the centre of the stem (sometimes only in the vicinity of a leaf insertion, sometimes throughout their course) than do the leaf-traces. In reality, as there is no interfascicular cambium in the normal sense, one might just as well reverse the form of statement and say that the traces had adopted a cortical position as leaf-traces have often done elsewhere. The meristematic cylinder of these stems not being equivalent to a normal cambium, the distinction between cortex and pith is quite arbitrary. So far as most of these plants are concerned, the notion of medullary bundles seems to have arisen partly because the traces are multiple, and in a transverse section the traces, if they outnumber the stem bundles, may very well appear as the 'normal' vascular cylinder. Only in the genus *Beta* does Wilson produce any evidence of more fundamental modification; a diagram which he copied from Fron (1899) shows bundles crossing over tangentially in a manner which could not be reproduced by any plastic deformation of a cylindrical system.

Many of the Chenopodiaceae are markedly xerophytic. In some of these (*Salsola*, *Suaeda*) the leaf, although small, remains the principal organ of photosynthesis, but is more or less succulent. In others (*Salicornia*, etc.) a more extreme condition is found in which the photosynthetic function has largely been transferred to the stem; each internode has a palisade layer in its outer cortex and is considerably swollen, while the leaves, which in these species are invariably decussate, may be reduced to inconspicuous bulges only just large enough to shelter the axillary structures. The plant is then a stem succulent of characteristic jointed appearance. The vascular systems of several species of *Salsola* and *Suaeda* have been examined by Fahn & Broido (1963); disregarding certain points of detail it appears that species with opposite leaves conform to the pattern shown in Fig. 101 (*Salsola soda*, *S. longifolia*, etc.) or to the more condensed scheme of Fig. 102 (*Salsola rosmarinus*), while species with spiral phyllotaxy are nearly as shown in Fig. 103 (*Suaeda palaestina*). These are very ordinary vascular patterns for small dicotyledons without stipules. In the Chenopodiaceae with succulent stems, however, additional features are found. The stem here contains the central vascular system and also a system of cortical vascularization which traverses the internode at the boundary where the outer (photosynthetic palisade) layer of the cortex adjoins the less specialized parenchymatous inner cortex. When the cortical vascularization is seen through the outer surface of the internode it is found to constitute a dense reticulum of a kind which is common in leaves but very different from the usual vascular pattern of an internode. Fahn & Arzee (1959) examined a number of species without finding any particular novelty in the central vascular system; *Anabasis* and *Haloxylon* conformed to Fig. 102, while *Arthrocnemum*, *Halocnemum*, and *Salicornia* followed the even simpler pattern of Fig. 104. The cortical vascularization appeared to be a simple addition to the normal scheme. In all cases the cortical system is connected to the main stele principally at the upper end of the internode. On its way into the reduced leaf, the leaf-trace divides into three, and the lateral arms, instead of entering the leaf, lead into the cortical network of the internode below. As this cortical system, unlike that of most stems in which cortical bundles occur, is not continued from one internode to the next, any photosynthetic product which enters a cortical strand will ordinarily have to move upwards to enter the central vascular system. Only in *Anabasis*, where the internodal part of the leaf-trace makes some supplementary and more direct contributions to the cortical reticulum, is there any significant variation in the plan.

On the whole, therefore, the vascular construction of the Chenopodiaceae involves surprisingly little modification of the basic pattern of stem bundles and traces. A more fundamental change can be observed in *Ricinus*, where there may be as many as twenty-three traces to a leaf in almost as many gaps, and where the number of bundles in an internodal section may be sixty or more. There is here at every node a network of vascular strands transversing the pith, not quite horizontally but in a sloping plane related to the levels at which the leaf-traces leave the cylinder. This system has been described and illustrated by Reynolds (1942), whose account is unfortunately in a descriptive convention which leads automatically to the statement that the bundles in the septa of the stem are traces (p. 156). It seems clear, however, that none of the bundles crossing the pith is a trace in the sense in which the term is used in this book. The horizontal nodal complex is essentially a new development, and there is no obvious advantage in terminology which obscures its novelty. A very similar plate-like transverse net of vascular cross-connections is found also in *Peperomia*, where however the internodal bundles, instead of forming a cylinder, are scattered in the manner of monocotyledons. In *Peperomia*, unlike most other dicotyledons, the leaf-trace bundles have virtually no free course in the stem. At each node the internodal bundles largely surrender their identities in the transverse net, and it is from the edges of the net that bundles are drawn off to serve the leaf. The vascular plan is substantially the same for species in which each node has a single large leaf as it is for those with smaller leaves in whorls.

We may provisionally associate with *Ricinus* and *Peperomia* the stem of *Piper*, which has long been noted for the possession of medullary strands inside a normal ring with continuous secondary tissues. An account by Hoffstadt (1916) is based on the concept of the descending trace. In *P. methysticum* there are two rings of medullary bundles, this arrangement being attributed to the fact that traces when followed downwards run for one internode in the stele, move into the pith and run for one internode as outer medullary strands, then move still further in and run for yet another internode as members of the inner medullary group. This condition contrasts with that of *P. umbellatum*, which has only one ring of medullary bundles, and in which 'the bundles, instead of transversing two nodes as internal bundles before fusing with the next group, traverse but one'. There is no reason to question the accuracy of these statements as a description of the order in which the various parts of the vascular system develop. They tend however to relegate some parts of the system unwarrantably to a position of subordinate rank. The arrangement of the vascular connections between the rings of bundles was not satisfactorily elucidated, and was perhaps regarded, under the influence of the descriptive style adopted, as a matter of no great consequence.

The first effective study of the vascular connections in a typical monocotyledon was perhaps that of Evans (1928) who worked with *Zea* and developed a technique for injecting methylene blue into a selected few of the larger bundles of an internode. He then removed the parenchyma of the stem by bacterial action, and was able to trace the course of the stained strands through the nodal complex. This procedure seems to leave no possibility of establishing any simple rule about the behaviour of any individual strand; such generalizations as may be attainable can only be of a statistical character. It appears from the work of Evans that on approaching the node the smaller peripheral bundles of the internode enter into a network in which their individuality is entirely lost. So far as these strands are concerned it is meaningless to speak of an identity between a bundle below the node and one above. The larger bundles deeper in the stem, however, can often, though not invariably, be followed through, and earlier workers had claimed that they very often passed unchanged through the surrounding confusion. It was supposed that one of these larger strands might commonly run through five or six nodes without being connected to anything else. Evans, with every show of reason, rejects this view as based upon inadequate standards of dissection. In reality the node contains not merely one but several layers of small strands crossing the stem horizontally, and at various places there are also fine vertical connections between these horizontal nets. It is quite true that the larger vertical strands appear to run straight through, and that a horizontal connection, on approaching such a strand, will often form a loop round it as though there had been some reaction of avoidance. There are however enough connections (most of them admittedly of small calibre) running into or out of the vertical bundles to make it rather exceptional, though not impossible, for any of these to retain exactly the same constitution through any significant number of nodes.

The vascular situation in the stems of the larger monocotyledons is so complex as to defy analysis by unaided visual comparison of sections. Zimmermann & Tomlinson (1968) have made significant progress by the use of an 'optical shuttle'. Essentially this is a switch-over device between two identical microscopes, both serving a common ciné camera. Serial transverse sections are mounted one per slide and are brought to the two microscopes alternately. By means of the shuttle it is possible to ensure that successive slides are accurately registered in the focal plane of the camera. Single exposures are made, and the end-product is a film which, upon projection, gives an impression of a journey through the stem, the bifurcations, displacements, and anastomoses of the bundles appearing on the screen in motion. The detailed analyses which have been carried out in this way have tended to confirm what has been known in outline for a very long time, namely that in these plants the larger strands which enter the leaf have a characteristic course, moving slowly from the periphery of the stem to positions near the centre, and then turning sharply outwards again to reach the leaf, as in Fig. 105. There can be little doubt that this behaviour is related to the large diameter which the stem of these plants attains in its primary expansion. Fig. 105 will indicate how the inward curve is formed in the vicinity of the apex. The intensive comparison of sections has however added much that is new. It has become clear that the strands often perform curious evolutions in the region of the bend. In palms the bundles in the central area engage in a kind of spiral swirl which is related to the phyllotaxy. In *Prionium* (Juncaceae) there is no general rearrangement, but each individual bundle at its nearest approach to the stem centre executes a smart 360° twist. It has been discovered also that the vascular system contains numerous cross-connections which the early observers overlooked. Typically the leaf-trace in its sloping outward passage through the peripheral zone of the stem gives off several branches which either continue vertically upwards as additions to the bundle-population of that area, or join with vertically-running strands already present. Furthermore these bundles which spring from the outward-moving leaf-trace tend to be histologically distinct, with a specialized type of small-celled phloem which does not occur elsewhere in the stem. Once the significance of this peculiarity has been recognized, these connecting bundles can be detected in single transverse sections. The ability of a leaf-trace to establish such connections is apparently related to the earliness of its own appearance. Young leaves in positions such as B in Fig. 105 are still differentiating additional vascular strands, but these will not have the full inward bend of the first-formed traces, and will have less opportunity to form cross-connections in the stem, indeed in many cases the last-formed foliar strands fail to make contact with any other bundle, and end blindly when traced downwards in the stem cortex. In *Prionium* about 30% of the bundles of the leaf are thought to have no downward connection; clearly the plant may here be approaching a natural limit to the possibilities of evolutionary progress in this particular direction.

The vascular strands in a petiole or rachis, though very variously disposed in transverse section, form in general a system of parallel bundles with few cross-connections, a state of affairs rather similar to that which prevails in the internode of a stem. It is however a common feature of leaves in which separate strands exist in the petiole that these should be linked by a massive collar or bridge in the leaf-base. This collar is often situated at the upper end of any stipular or sheathing region which may be recognizable. In many compound leaves a similar type of cross-connection may be associated with the attachment of each pinna. Some measure of analogy can therefore be said to exist between the vascular situation in petiole and rachis and that in a stem. The comparison cannot however be pursued to any great refinement of detail. The bifacial symmetry of the leaf and its specialized mode of growth ensure that the nodule (the region of attachment of a leaflet) can never fully repeat the characteristic features of a node. Concerning the vascular connections between the midrib and the principal lateral veins of simple leaves we appear at present to be almost totally uninformed, but the larger examples certainly have room for a good deal of anatomical complication.

The venation pattern of the lamina can in most cases be treated as two-dimensional and can be observed without significant technical difficulty. Venation patterns are also strikingly diverse, but attempts to study them comparatively have not yet led to any very distinguished result. The difficulty, fundamentally, is one of analysis. In examining a set of drawings of venations, as for example the fine illustrations of Araceae by Ertl (1932) it is easy to see that the species differ, and that pictorial aids can be of value in such problems of identification as may arise in ecological or palaeobotanical investigations. Comparisons of this type are however usually conducted at a level of general subjective impression which does not demand precise geometrical thought on the part of the observer. As soon as we try to specify the exact nature of differences, or to recognize strictly homologous parts in different systems, our troubles begin. It may be helpful to separate four classes of phenomena:

I. It is sometimes possible to delimit areas in which the veins display a common sweeping curvature suggestive of some controlling field of force. In most such cases, perhaps in all, the venation pattern appears to be related to a polarization of lamina growth in one dominant direction during the period of vascular differentiation. The parallel venation of a grass leaf is the simplest case of all, where growth is for a long time almost exclusively longitudinal. Marginal growth radially outwards from a centre produces a fan-shaped area with radiating veins, as in the beautiful photographs of *Regnellidium* published by Pray (1962). Growth laterally outwards from a central midrib produces the characteristic leaf of the Scitamineae. These are cases in which a single law of vein orientation prevails over virtually the whole leaf surface and the venation of the lamina is closely related to its general shape because both are consequences of the same simple pattern of growth. Concordance of vein directions is however often more localized; we may instance the special conditions in the basal lobes of some sagittate leaves or between the lateral ribs in *Quiina* (Foster, 1950).

II. It is possible to base a few clear-cut morphological distinctions upon the behaviour of the principal vascular strands as they approach the margin. Thus in the vascularization of a more or less elliptical lamina we can recognize the types shown in Figs. 106–108, and can draw attention to the fact that the pattern of Fig. 106 is particularly characteristic of *Fagus*, *Corylus*, and related genera, and is found in Trifolieae but not in other legumes, while Melastomaceae and many monocotyledons are nearer to Fig. 108. Once the facts are stated, however, it is not clear how we can proceed further. The same limitation applies to Pray's observation that the radially dichotomizing veins of *Regnellidium* are joined by a submarginal strand, whereas the otherwise similar system of *Ginkgo* has none.

III. Some attempts have been made to base taxonomic distinctions upon the purely local quality of the minor venation. The most successful exercise of this kind is probably that of Dede (1962) on the admittedly specialized case of the Rutaceae. In these plants the leaf contains globular secretory cavities. Dede has examined members of eighty genera and revealed significant differences in the extent to which the vascular pattern is affected by the presence of the glands. The situation which is regarded as primitive, and which is still probably the commonest condition in the family, is that in which the rather numerous glands are small enough to lie between the laminar venation and the epidermis, so that in surface view there is no correlation between the two tissue systems; glands may occur in the areoles but are just as likely to lie over a minor vein. In types which are considered to show an evolutionary advance, the glands become progressively fewer and larger. In some cases they merely occupy areoles which are not significantly modified from the vascular point of view, but the sequence ends with species in which each gland is surrounded either by a ring-bundle of regular form or by a less coherent but quite distinctive complex of radially-arranged strands. Dede calls this latter configuration a 'sunburst', and illustrates various subsidiary categories. In most families the minor venation of the leaf will not offer such obvious points of distinction as are presented by Rutaceae, but it is clear that much remains to be done. Thus Ertl observed that some Araceae have a great many free vascular endings, while others have hardly any. Lems (1964), in a study of Ericaceae, was able to make some use of the form of the vein endings, which may be pointed, bulbiform, or splayed into knobby branches.

IV. There are some leaves in which the vascular system of the lamina does not lie in a single surface. In Araceae Ertl describes vascular crossovers in which two strands overlie one another at an angle. This can occur in a lamina which is not conspicuously thick, and one strand is appreciably deflected towards the epidermis as it passes over the other. Sometimes, but by no means in every case, there is intercommunication at the crossing. These crossovers are unknown in some species, occur sporadically in others, but are in some a regular feature, so that a leaf may in effect have two vascular networks superimposed. The development of the venation upon more than one level naturally tends to assume greater prominence where the leaf is thick. A good example may be seen in the work of Carlquist (1957) on *Argyroxiphium* and *Wilkesia*. These are Compositae, endemic to Hawaii, with the general growth-habit of *Agave* and *Yucca*, and resembling the former in being commonly monocarpic. In their thick sword-shaped leaves there are reticulate venation systems at three distinct levels. The one in the middle level is the 'normal' venation derived directly from the trace bundles. The system near the upper surface springs from the 'normal' system at some little distance from the stem, and is consequently not represented in the leaf-base. The abaxial system, however, running near the lower epidermis, is a largely independent development. It appears in tissue derived from a special meristem in which division walls are parallel to the lower surface, and which adds substantially to the thickness of the leaf. The abaxial system connects with the other veins principally at the margins and tip of the leaf, and has no direct continuation in the basal direction. This condition of relative isolation is curiously similar to that reported in large monocotyledons (see p. 178) and will bear comparison also with the cortical system in the stems of some Chenopodiaceae (see p. 174).

The nature of the minor venation of reticulately veined leaves has been the subject of some dispute. It was claimed by Slade (1957) that the vein-endings which so characteristically occur within the areoles or vein-islets arose by breakage: that is to say, the venation at an early stage consisted of closed meshes only, and that some of the strands, failing to keep pace with the general growth of the lamina, were pulled apart to create blind endings. Upon this view, one might expect to find vein-endings pointing towards one another in pairs, but this is a very rare configuration. It now seems reasonably clear that Slade was mistaken, and that in ordinary cases the whole system is formed by progressive procambial development, differentiation spreading in an orderly manner from the larger strands to the smaller ones, and so to the terminations. Pray (1963) has given some interesting details, showing how important veins may be flanked by 'non-areolate' areas in which there is for considerable distances no branch from the large vein, and finding also large areoles in which there is a complex branched inner system, sometimes with closed meshes of its own. The effectiveness of the vascular network is a matter of obvious importance, and several kinds of measurement can be taken. Von Zalenski (1902) determined for a considerable number of species (mostly herbaceous) the total length of vein per unit area of lamina, obtaining generally values in the range 50–100 cm per cm^2, with lower values in hydrophytes and higher ones in particularly arid conditions. Plymale & Wylie (1944), applying the same method to deciduous trees and shrubs, obtained figures which lay mostly in the upper part of the range of von Zalenski's observations, around 100 cm per cm^2, and further made a distinction between major veins (those associated with a conspicuous rib on the lower surface) and minor ones running entirely within the normal thickness of the lamina. They found a relatively constant 95% of the total length of vein to consist of the minor strands. They also considered what proportion of the area of the lamina should be regarded as occupied by veins to the exclusion of photosynthetic tissue and obtained figures of 20–25%, pointing out that rates of photosynthesis expressed in terms of total leaf-area would underestimate the activity of the mesophyll.

Plymale & Wylie went on to perform a series of experiments in which large portions of lamina were partly isolated by cuts, the incisions being so arranged as to leave narrow bridges of tissue. Supporting threads were provided to secure the mutilated leaf in position, and transpiration was allowed to proceed normally. Any failure of the remaining vascular network to supply sufficient water to the mesophyll is in these conditions quickly revealed by the death of tissue. Such experiments consistently demonstrate the ability of the venation to carry a very large conductive overload in almost any direction. A narrow bridge containing only very small strands can convey enough water to sustain a large area of lamina beyond the bridge. The only type of bridge which is relatively ineffective is that shown in Fig. 109 (see p. 184), where a portion of a major strand lies across the path along which water is required to move. The widespread death of tissue in experiments of this particular form is presumably related to Pray's observation that a major strand may have only sparse direct connections with the minor network on either side of it. The demonstration of excess conductive capacity in the vascular system applies only to water movement; conditions relating to carbohydrate flow may be quite different.

In a further paper Wylie (1946) attempted to assess the potentialities of the kinds of cell which make up the non-vascular tissues of the lamina in terms of their suitability for water-conduction in various directions. He argued that an epidermal cell, being in close contact with other epidermal cells but in less intimate contact with the internal tissues, is most easily able to pass materials across, rather than through, the lamina. A palisade cell, on the other hand, will more readily convey substances parallel to its own longer axis, that is perpendicular to the leaf-surface. The mesophyll between the finest veins must in any case be served by conduction through the non-vascular tissues, and it may plausibly be argued that the maximum permissible distance between a palisade cell and the nearest vascular element will be greater where the parenchymatous layers of the lamina are best suited to paradermal conduction. On this basis veins might be more widely spaced in leaves with thick epidermis and spongy mesophyll, closer together in leaves with a higher proportion of palisade. Wylie made observations on a range of species, and from his results we can calculate Fig. 110. What has here been called 'average minimum spacing of veins' is a measurement not perhaps so accurately standardized as one would like, but there can be no doubt that the expected relationship is a real one. That it arises from considerations of conductive efficiency is of course much less certain; it would not be difficult to devise alternative explanations concerned with the developmental mechanics of the lamina. Philpott (1953), in a large comparative study of *Ficus* species, has confirmed the relationship between vein spacing and tissue continuity.

Any attempt to put the study of reticulate venation on a quantitative basis will obviously encounter problems of the utmost complexity, and it is natural therefore in the present undeveloped state of the subject that attention should be concentrated on the simplest examples obtainable. There have been several investigations on the occurrence of vascular anastomoses in systems where anastomosis is a relatively uncommon feature, so that each occurrence can be treated as an isolated and independent event. Arnott & Tucker (1963) chose the petal of *Ranunculus*, in which the vascular system is basically a succession of dichotomies. So far as the frequency of cross-connections is concerned, we can recover from their account the numbers set out in the second column of Table 16. The commonest class of petal appears to be that in which there is one, but only one, event of vascular fusion. Petals exist, however, with no anastomoses at all, or with as many as eight. In these circumstances the question of contagion arises: does the occurrence of an anastomosis in a petal increase or diminish the likelihood that that petal will have further anastomoses also? The original authors did not investigate this matter, but we may do so by calculation from their counts, as at the right of Table 16. The numerators and denominators of the fractions are obtained by summation. For instance, there are altogether 217 petals with at least three anastomoses apiece, and of these ninety-four, or 43%, also possess a fourth anastomosis. The predominant declining trend of the probabilities in the last column of the table suggests that each anastomosis added to the system progressively reduces the prospects for the development of others.

The anastomoses which appear in a vascular system of this nature are not all of the same morphological status, but can be sorted into several categories. Arnott & Tucker adopted for this purpose a classification which is perfectly usable but which is open to criticism because it tries to take into account several characters at the same time. Thus of the six

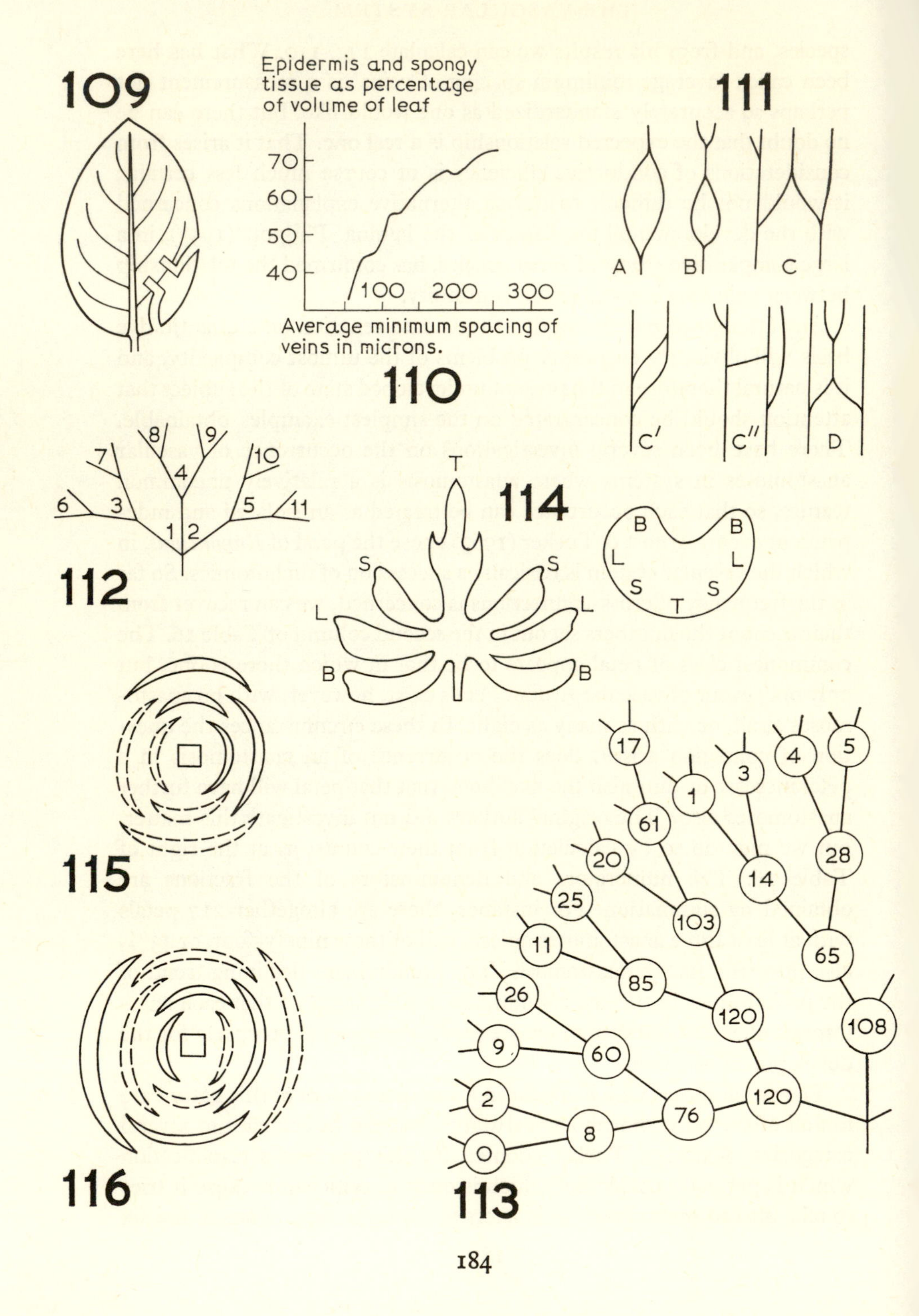
109
Epidermis and spongy tissue as percentage of volume of leaf
70
60
50
40
100
200
300
Average minimum spacing of veins in microns.
110
111
A
B
C
C′
C″
D
112
8
9
7
10
4
6
3
5
11
1
2
114
T
S
S
L
L
B
B
B
B
L
L
S
S
T
115
116
17
5
4
3
1
61
28
20
14
25
103
65
11
85
26
120
108
9
60
2
120
76
8
0
113

TABLE 16

Distribution of vascular anastomoses among 1218 *Ranunculus* petals (data of Arnott & Tucker, 1963)

Observed number of anastomoses	Number of petals in sample	Probability of occurrence of at least one additional anastomosis
0	372	$\frac{846}{1218} = 0{\cdot}70$
1	378	$\frac{468}{846} = 0{\cdot}55$
2	251	$\frac{217}{468} = 0{\cdot}46$
3	123	$\frac{94}{217} = 0{\cdot}43$
4	57	$\frac{37}{94} = 0{\cdot}39$
5	19	$\frac{18}{37} = 0{\cdot}49$
6	11	$\frac{7}{18} = 0{\cdot}39$
7	5	$\frac{2}{7} = 0{\cdot}29$
8	2	

FIGURES 109–116. *Vascular structure.*

109. Experiment by Plymale & Wylie (1944) in which a portion of a leaf remains connected only by a narrow bridge which is crossed at right-angles by a major vein. **110.** Graph calculated from observations by Wylie (1946) showing relationship between spacing of the finest leaf veins and the histological constitution of the intervening lamina. **111.** Types of anastomoses distinguished by Arnott & Tucker (1963). **112.** Beginning of a system for numbering the dichotomies in a *Ranunculus* petal (Arnott & Tucker, 1963, 1964). **113.** Frequencies of occurrence, in a sample of 120 specimens, of particular dichotomies in the left half of a petal. **114.** Correspondence observed by Yin (1941) in *Carica*; the vascular supply to any lobe of the leaf passes mainly through the similarly lettered part of the transverse section of the petiole. **115.** Two-sided system of deblading used by Rossetter & Jacobs (1953) and Jacobs (1958). Diagram is plan view of *Coleus* shoot; leaves shown dotted are cut away to leave only the petiole standing. **116.** Spiral system of deblading.

types shown in Fig. 111, B differs from A, just as D differs from C, only in the bifurcation, or failure to bifurcate, of the compound strand arising from the anastomosis. That this is a sufficient reason for separating the anastomoses into entirely different classes is not immediately obvious. Again it may be argued that the type C′, in which a small loop is formed by the fusion of strands closely related in origin, has more in common with A and B than with C and C″ which are associated with it by the scheme of lettering. Also, as the authors themselves admit, C″ is not truly distinct from C, differing from it only in that one of the anastomosing bundles is 'more or less' at right angles to the other strands. Altogether the scheme embodies a considerable number of decisions which, though not necessarily indefensible, were of a quite arbitrary character. The advantage to be gained by giving separate recognition to classes like A and B, which occur so rarely that their frequencies cannot be determined, is in any case very slight.

In the 1218 petals of Table 16, there were 1665 anastomoses. We will divide these into two types only, those which united strands coming from a near-by common source so as to enclose a small loop (types A, B, and C′ of Fig. 111) and those which joined strands of less closely related origin (types C, C″, and D). There were 468 fusions of closely related shanks, 1197 fusions of more independent shanks. The two types are differently distributed over the area of the petal; we show this by two charts (Table 17). Here the petal is divided into nine parts shown as squares; Arnott & Tucker had fifteen divisions, with five longitudinal strips of which we are combining the central three. The number of anastomoses found in each division is shown on a percentage basis. Evidently the fusion of independent shanks is concentrated in the middle of the petal, whereas the loop-forming type of anastomosis is sharply characteristic of the apical and lateral shoulders. In these areas the 'independent shank' type constitutes only about 56% of the total anastomoses occurring, whereas in the centre 99% of anastomoses are of this class.

It is shown in the original publication that the number of anastomoses per petal is larger in long petals than in short ones and also that petals which are prominently two-lobed at the apex have more anastomoses than those which are not. This information, however, is less satisfying than it first appears. Lobed petals are on average larger than unlobed ones, and it has not been demonstrated that lobing has any independent effect (independent, that is, of the greater length which goes with it).

TABLE 17

Topographical distribution of two types of vascular anastomosis in *Ranunculus* petals. Petal divided into nine regions, frequencies as percentages. Data of Arnott & Tucker (1963), same sample as Table 16

Anastomoses between closely related shanks				
		Apex of petal		
Left side	31%	2%	35%	Right side
	14%	1%	14%	
	1%	0%	1%	
		Base of petal		
Anastomoses between independent shanks				
		Apex of petal		
Left side	15%	13%	17%	Right side
	13%	27%	12%	
	1%	1%	0%	
		Base of petal		

Furthermore the measurement of petal length is rather too remote from the local conditions in the lamina. What we really need to know is the relationship between anastomosis-frequency and the number of locally available bundles between which anastomoses could potentially occur. Arnott & Tucker's published tracings suggest that even a small petal may have several fusions if enough dichotomies occur in it, while a large and deeply lobed petal which is sparsely vascularized will have few if any anastomoses. A survey based on dichotomy-counts would be preferable to one based on measurements of length or on the partly subjective recognition of a lobed outline.

In a later publication (Arnott & Tucker, 1964) all the possible sites at which a dichotomy could occur were identified by a scheme of numbering (Fig. 112); the division of the main petal bundle into three is represented by the two almost coincident dichotomies 1 and 2, while numbers 3, 5, 7, and 10 are also constantly present. The frequency of occurrence of a dichotomy at any other position can be determined for petals of any specified size-range. In general the probability that a dichotomy will exist at any point will increase as the petal becomes larger. Arnott & Tucker's tables, however, show very clearly that some places are inherently more favourable to vascular bifurcation than others. As the petal is approximately symmetrical it will be sufficient for us to show (Fig. 113) the frequencies of occurrence of dichotomies relating to the left-hand half; the numbers here are not percentages but refer to a sample of 120 specimens. A pattern can be seen to unfold as the process of dichotomy proceeds. Throughout the left-hand arm of the system, it is a rule that the right-hand shank of a dichotomy, that is to say the one nearer to the midline of the petal, is more likely to undergo dichotomy of the next order than is its partner. We have such comparisons as 120 with 76, 103 with 85, and so on. The left-hand branch of the central arm, within its more restricted compass, displays the same principle at work, and the right side of the petal follows the same rule that the more nearly median shank of any dichotomy shall be more fully developed. Extreme conditions are found at the far left where a zero is entered. This dichotomy, which is number 24 in the scheme of Fig. 112, and its counterpart in the right half (which is number 47) are the only dichotomies of their rank not to have been observed. On the other hand, of the dichotomies of the next order only two have been seen; they are numbers 63 and 80, each at the end of a long succession of 'inner' shanks.

In a tabulation of this kind we may see a possible starting-point for the development of a computable model. If we draw in the first instance only those dichotomies which are of universal occurrence in the sample (120 in Fig. 113) we obtain a basic framework. Addition of those which satisfy some slightly lower standard of frequency, say 100, will cause our diagram to grow a little, and this may progressively continue until the rarest dichotomies are added last of all. An animated film made in this way would present a naturalistic impression of spreading vascular differentiation. How closely this would conform to real ontogenetic sequences is not clear. We must not lose sight of the fact that Arnott & Tucker were comparing inherently large and inherently small petals, not older and younger ones. For the time being, however, their work appears to constitute the furthest advance in this direction. Among these very simple, essentially dichotomous systems of venation, some interspecific comparisons can already be attempted. For example Arnott (1959) has given some tables for the occurrence of anastomoses in the leaf of *Ginkgo*, from which it is possible to carry out a calculation exactly parallel with that shown in Table 16. The results, however, are very different, with successive frequencies of 10%, 45%, 30%, 43%; here there is strong evidence that the comparatively rare occurrence of a first anastomosis in a leaf is markedly favourable to the appearance of others also, which is the opposite of the situation in *Ranunculus* petals. In *Ginkgo*, the frequency of anastomoses is greater in long than in short shoots, which must present opportunities for physiological investigation.

As the arrangement of vascular connections must often determine the direction of movement of physiologically active substances, it is likely at times to govern the way in which a plant responds to natural or experimental conditions. An example can be seen in the observations of Yin (1941) on the leaf of *Carica*. A lobed lamina (Fig. 114) is here connected to its petiole in such a way that the main communications with each lobe are concentrated in a particular part of the petiole cross-section (indicated by corresponding letters in the diagram). In natural conditions the leaf executes a daily cycle of movements, rising to a high position at daybreak, dropping rapidly in the evening, and showing a partial recovery before midnight. These effects are due to growth curvatures in the petiole. There is no pulvinus. A debladed petiole neither grows nor curves, whereas a debladed petiole supplied with auxin grows without curvature. It is reasonable in these circumstances to seek an explanation for the nyctinastic cycle in differential production of auxin from various parts of the blade. When a plant is kept in continuous light the leaves soon come to rest. Application of auxin paste to lobes B and L will then cause the petiole to curve down, whereas a similar application to lobe T will cause it to bend up. The final stage of Yin's investigation was the examination by an auxin assay technique of leaves kept under normal alternation of light and dark. This revealed the existence in each lobe of a regular cycle of changes in auxin concentration, the cycles in the terminal and lateral lobes respectively being out of phase by an amount sufficient to account for the observed deflections of the petiole. The operation of this mechanism depends on a vascular situation which appropriately localizes the differently timed releases of auxin within the transverse section of the petiole.

Physiologists who have had occasion to consider the relationship between one leaf and its neighbours have often shown an awareness of the possibility that much might depend upon the nature of the vascular connections within the stem. Thus Steward (1954) in studying the movement of mineral nutrients in the shoot of *Cucurbita* was led to the belief that in early life each leaf passes through a phase of very active absorption, during which it may draw salts from older leaves, as well as directly from the roots. Steward viewed this transfer phenomenon as primarily a flow through a specific stem bundle (one of a set of five) connecting an older leaf with a younger one five internodes higher. Even though this interpretation seems to disregard the possible significance of some vascular cross-connections in the stem, it may well be true that the most direct route is the one carrying the great majority of the traffic. Independent evidence bearing on the importance of vascular pattern in relation to hormone flow is furnished by the work of Rossetter & Jacobs (1953) and Jacobs (1958). The effect examined was the ability of intact *Coleus* leaves to accelerate the abscission of debladed petioles standing near them on the stem. Speed of abscission was the measurement to be recorded, and as the phyllotaxy is decussate several patterns of leaf-deblading were possible. To remove all the leaf-blades on two (adjacent) faces of the stem and leave intact all the leaves of the other two (adjacent) faces (Fig. 115) is called two-sided deblading, while the pattern shown in Fig. 116 is spiral deblading. The two are equivalent as regards the total loss of tissue, just half the laminar surface of the shoot in each case. Spiral deblading, however, gives quicker abscission of the debladed petioles than two-sided deblading, and this is attributed to the difference in conditions of hormone flow. Assuming, as other forms of experiment strongly suggest, that speedy abscission is caused by hormone from a surviving lamina, the vascular connections are such that the flow will pass most readily between leaves in the same vertical row; in spiral deblading vertical flow is all that is required, but in two-sided deblading the laminar hormone can reach the debladed petioles only by a less direct and consequently less efficient route.

Schwartz (1928), also working with *Coleus*, introduced questions of vascular continuity into his discussion of a particular type of leaf variegation. In his plants a young leaf is green, but a yellow discoloration subsequently spreads outwards from the centre towards the margins. There is evidence that this is not the result of a chimaeral structure, but that a substance capable of decomposing chlorophyll is formed in the main veins and spreads outwards from them. An incision running alongside the midrib (Fig. 117, see p. 191), if made early enough, will locally prevent this spread. The active substance, however, appears to be locally produced (perhaps in the bundle sheath?) rather than imported by the midrib, because severance of the midrib (Fig. 118) does not disturb the normal pattern. Schwartz was able to obtain a shoot with the colour distribution of Fig. 119, and argued that this was a product of the regularity of the vascular connections at the nodes. This claim is not necessarily sound and there is something paradoxical in invoking a vascular explanation for the distribution of a substance which appears not to travel through the vascular system. There is no doubt, however, that the nature of the vascular connections ought always to be taken into account in dealing with such problems of distribution.

It is evidently desirable to investigate the genetical basis of vascular differences, but research workers are unlikely to be attracted by the prospect of genetical experiments where the results are so inaccessible as the vascular pattern in a typical vegetative stem. Nor can the obstacle be surmounted by a simple decision to work on foliar venation; admittedly the purely technical difficulty of ascertaining the effect of an experimental hybridization would then be negligible, but the vascularization of the leaf is so complex, and so different from that of the stem, that difficulties of interpretation and analysis must largely replace the difficulties of observation which characterize the stem. A reasonable compromise is perhaps to be found in the examination of the floral receptacle. Here a large amount of vascular rearrangement is concentrated in a short distance, reducing the work of sectioning to tolerable

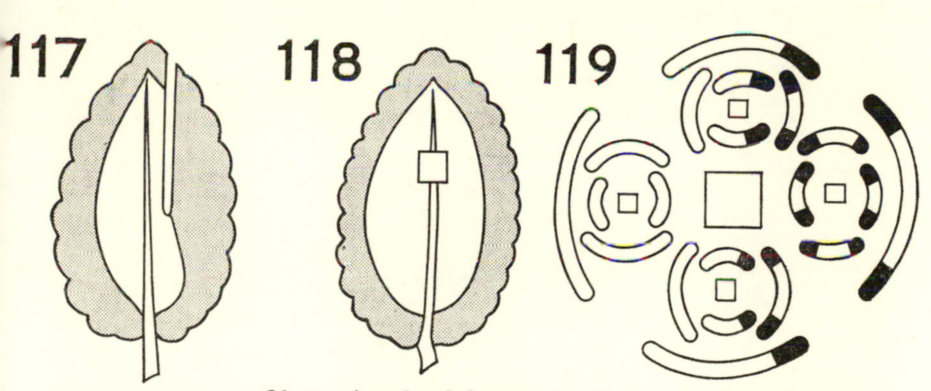

FIGURES 117–119. *Observations by Schwartz (1928) on a variegated form of* Coleus, *in which the leaf is progressively decolorized by a chlorophyll-destroying substance spreading from the midrib.*

117. Leaf with an incision parallel to the midrib shows partial preservation of the chlorophyll (green areas stippled). **118.** Leaf with severed midrib shows normal destruction of chlorophyll beyond the wound. **119.** Plan view of a sectorial chimaera shows chlorophyll (here represented black) on one side only. The extreme accuracy with which this pattern was continued in subsequent growth was attributed to the nature of the vascular connections.

proportions, while the range of inter-specimen variation within a single genotype will often be pleasingly restricted. An indication of the possibilities can be seen in a study by Hillson (1963) of some hybrids between *Mentha spicata* and *M. aquatica*. In these plants the vascular strands of the pedicel join into a continuous cylindrical stele in the lower part of the receptacle. In *M. aquatica* this cylinder terminates in an almost uniform transverse upper rim, from which all the petal and stamen traces arise at a common level, whereas in *M. spicata* the stelar cylinder is taller but has a markedly uneven upper rim, and is deeply cleft on its posterior side. There are differences also in the vascularization of the ovary and other details. Interspecific hybrids, while showing in some respects a mixture of characters, are much nearer to *M. spicata*. The example is perhaps not a specially informative one; it might be argued that the differences recorded are mainly consequential upon the difference between a species which is strongly zygomorphic and one that is less so. Certainly there is a serious doubt as to how many of the differences described by Hillson are truly independent of one another. It is satisfactory however to observe that dominance relationships can apply to vascular features, and there is hope that other material may prove amenable to genetical investigation.

Chapter Seven

The Shoot System in its Environment

The experimental study of the relationships between a shoot and the conditions of its environment has lain principally in the hands of plant physiologists, though ecologists and others have made important subsidiary contributions. The resulting coverage of the subject is remarkably uneven, and while a few selected topics have been pursued with the utmost tenacity, there are many others of no less biological interest upon which only scanty information is available. The reasons for this are partly historical; tropisms and other effects which were readily demonstrable with the limited technology of 50 or 100 years ago have had time to acquire an entrenched position in the literature and in public esteem, to the extent that it would be hard to find a living botanist whose ideas about sensitivity in plants had not been founded on his early acquaintance with classical results in geotropism and phototropism. There is also an important influence arising from the education and tastes of the people concerned; the average modern physiologist has been conditioned by his training to seek problems which can be dealt with by chemical manipulation and in general to avoid those which obviously call for the application to the plant of refined mechanical instrumentation. Such preferences, in so far as their possessors pause to think about them at all, naturally tend to be protected by various types of rationalization. A man who is accustomed to think chemically, and who is unfamiliar with measurements of friction, can always perceive quite clearly that the growth of a rhizome is regulated by hormones, and that any detailed consideration of the mechanical resistance of the soil would be contrary to the best interests of science.

When we survey the growth of shoots in nature the strongly selective and rather artificial character of the picture drawn from physiological enquiries soon becomes apparent. Considering the attention which has been devoted to simple tropic responses in the laboratory, it is surprising to find how little of its time and energy the average wild shoot will devote to exhibitions of phototropism and geotropism. It is evident also that environmental influences are rarely of the elementary simplicity which one seeks to attain in laboratory experiments. Astronomical considerations, for instance, ensure that static unilateral illumination will be uncommon. Generally the illuminant swings round the plant in an arc; it is not entirely frivolous to argue that in photoperiodic experiments the main light period should come from lamps to the south of the plant and that the 'midnight flash' which is so often applied ought to be administered from the north. We have both a theoretical expectation that 'photoangular' effects may arise, and a body of evidence (see p. 203) for their actual occurrence, yet the laboratory investigation of phototropism has hardly started to attack the problem of moving light sources. Many other environmental factors, such as wind, are inherently complex in their action. The force of gravity is perhaps the only feature of natural environments which has been truthfully represented in experimental work on a large scale; the difficulty is, of course, to get rid of it to obtain a control. It is desirable therefore to consider the responses of shoots to many stimuli which have not often formed the subject of experiments, and some of which will undoubtedly be very difficult to quantify. One aspect of such a survey ought to be a critical re-examination of phenomena which have in the past been referred to presumptive causes by writers who had too restricted a view of the possibilities. Some characteristic orientations of leaves, for example, have traditionally been attributed to phototropism and/or geotropism, not so much on direct experimental evidence as upon a general belief that these tropisms dominate the behaviour of shoot systems to an extent which virtually excludes any need to seek alternative explanations. There may well be cases in which a more balanced view would noticeably transfer the burden of proof.

It will not, however, be sufficient to consider in isolation any particular form of interaction between the shoot and its environment. Generally it is necessary to deal with long developmental sequences, in which different types of behaviour follow one another in an orderly succession. From the organizational standpoint what has to be examined is not so much the existence of a particular tropism or other response mechanism, as the part which may be played in the life of the plant by the emergence of a characteristic type of sensitivity at particular times and places. The contemplation of elementary effects such as tropisms, and possibly also the difficulty of clearly visualizing the relationships of successive events in the very slow tempo which so many plants display, have distracted attention to some extent from the occurrence in plants of behaviour patterns comparable in elaboration and general interest with those that attract attention in animals. The absence of a nervous system has undoubtedly contributed to the hesitation which many botanists have felt in this connection. In reality it is very doubtful, however, whether the behaviour of a flowering shoot or a rhizome system or a climbing plant is in any way simpler than that of an animal hunting its prey or building a nest. The processes are certainly slower, and a higher proportion of them involve changes in the bodily structure of the organism, and these considerations probably, on balance, make the study of plant behaviour more difficult. It does not follow that they make it less important or less interesting. Our task is therefore a double one, firstly to extend the knowledge of plant responses which have not yet been widely subjected to experimental enquiry, and secondly to distinguish the significant parts which the various elementary responses may play in complex chains of behaviour and development.

The literature of conventional plant physiology has tended to neglect many ecologically significant forms of mechanical stimulation. In the course of its life a shoot system is subjected to mechanical stresses arising from a variety of causes. Some are static gravitational loadings, others are transient forces such as may be imposed by the movement of surrounding air or water while a third category may be recognized in the stresses set up by the plant's own growth as it strives to overcome the resistance of external media such as soil, or the rigidity of protective structures such as seed coatings or bud-scales. The experimental investigation of responses to mechanical stress is a most difficult enterprise, and the progress which has been made is not comparable with that achieved in the much simpler study of tropisms and nastic movements. It is almost impossible to apply to a living plant a pure force of tension or compression; there must ordinarily be some measure of bending and (as for example by the attachment of a thread) some degree of localized contact stimulation. The speed with which a force is brought to bear must also be an important consideration; the instantaneous application of a heavy weight will generally have less relevance to ecological problems than the gradual building up of a similar stress from an initial low value over a period of hours or days. This desire to apply forces in a time-sequence which will bear some relationship to natural conditions raises complex problems of instrumental design.

It has been repeatedly demonstrated that the development of 'mechanical' tissues such as collenchyma can be enhanced by applied mechanical stimulation of the shoot. Thus Walker (1960) grew *Datura* seedlings in pots which stood on a shaker platform which was violently oscillated to and fro for nine hours each day. The treatment resulted in a marked increase in the thickness of collenchyma cell-walls as compared with unshaken control plants; this increase was, however, curiously slow to appear, most of it coming about from the twenty-sixth to the fortieth day of the shaking routine. The weakness of the enquiry lies in the complexity of the treatment. Walker's plants undoubtedly experienced repeated bending of the stem, and this may have been the principal operative stimulus, but his motor ran at such a speed as to subject the whole plant to considerable inertial stress; there must have been extreme over-stimulation of the perception mechanisms involved in geotropism, and this throws a doubt upon the true interpretation of the results. A more natural approach can be seen in the work of Venning (1949) who grew celery in an artificial wind and found that collenchyma formation in the petioles was greater than in control plants raised in calm air. The number of collenchyma patches in the petiole was unchanged, but their cross-sectional area was about 50% greater, apparently due to greater cell-expansion and greater thickening of wall, rather than to any increase in the number of cells. That more critical standards of experimentation can be reached is shown in the study by Himmell (1927) of the petiole of *Podophyllum*. The effect of bending was here investigated by means of a mechanical flexing device. The stout and normally erect petiole was bent repeatedly backwards and forwards in a single vertical plane. Instead of judging the effects by anatomical inspection, Himmell tested his specimens on a 'flectometer', a purpose-built instrument designed to measure the resistance of a petiole to bending. In the flectometer the petiole is clamped at its base and its tip is pulled sideways by an applied weight. It is necessary to be able to compare specimens of different lengths; the stiffness reading is therefore the weight required to displace the petiole tip by a standard amount for each unit of the petiole's length. By this means it was found that repeated flexing of a petiole brought about a 12% increase in stiffness in the plane of flexure, but a 40% increase in stiffness in a plane at right angles to the plane of flexure. This difference may to some extent reflect a change in shape; the petiole, which is normally cylindrical, becomes under repeated bending slightly elliptical, with its major axis running across the plane of flexure. Static loading of a vertical petiole with gradually applied weights up to several hundred grams (enough to slow down, but not to stop, the normal elongation growth), also produced an all-round increase in stiffness of about 18%. All these stiffness increases were attributed to cell-wall development rather than to enhanced turgidity, because they were maintained after cessation of the treatment, even when plants were kept for several days in darkness.

Himmel's most interesting work was concerned with the force which a *Podophyllum* petiole exerts in a geotropic response. In his experiment a petiole was laid horizontal, with its apical part stiffened by being enclosed in a glass tube, and tended to return to the vertical by geotropic curvature of its basal portion. As soon as the petiole tip rose by a small amount (less than 1 mm) from its starting-point it completed an electrical contact and energized a solenoid, causing a metal shot to be dropped into a counterpoised container suspended from the petiole. In this way the load which the petiole has to lift in order to show a geotropic response is progressively increased; every time the petiole begins to rise it is knocked down again by the (apparently quite violent) addition of another shot to its burden. A timing device records the dropping of each shot. Under this treatment the petioles were very generally found to show a second maximum in their pattern of geotropic behaviour. Thus a group of six petioles, in their first hour of shot-dropping, released altogether 102 shot, but in their second hour only 42. This decline did not continue, the figures for subsequent hours being 32, 59, 62, 54, 43. It seems, therefore, that there may be two mechanisms of geotropic response, one which acts quickly where there is no resistance to movement, and another which comes into action as a reserve when an obstacle has to be overcome. There is a close analogy here between the effects of gravitational stimulation and those of flexure, in that both stimuli cause increases of stiffness. It would therefore be reasonable to look, as nobody yet seems to have done, for a short-term response to flexure and for a build-up of resistance to flexure in two or more distinct stages. It is also impossible to overlook the fact that Himmell's bending machine pulled the petiole out of the vertical and therefore necessarily applied some gravitational component, though admittedly not in a constant direction.

Another aspect of mechanical stress is the application of fluid pressure; in view of the existence of high pressures in certain internal tissues the question is not without interest even for land plants, and its relevance to aquatics is obvious. Reports upon this matter are scanty and inconsistent so far as the higher pressures are concerned, but the responses of submerged aquatics to moderate excess pressure were successfully investigated by Ferling (1957), using *Potamogeton*, *Elodea*, and *Ranunculus circinatus*. Beyond a certain very moderate limit (a pressure of about three atmospheres absolute, maintained continuously for a few days) all these species showed a common pattern of symptoms. At lower levels of exposure there were consistent interspecific differences, with *Potamogeton densus* appreciably more sensitive than *Elodea*. Provided that the pressure treatment is sufficient, however, its exact nature seems to be of very little consequence. Pressures as high as 500 atmospheres were applied, admittedly for only a few hours, without evoking any effect not obtainable at a more sustained three atmospheres. Pressurization beyond the lowest levels is an experience from which a plant does not soon recover; it produces a semi-permanent conditioning. The dominant feature is a great retardation in the differentiation of new organs of all kinds. Leaves unfold very slowly, axillary buds fail to grow into branches and the emergence of adventitious roots is suppressed. Conversion of a vegetative apex to produce flower-buds is also markedly discouraged. In respect of all these processes there seems to be a principle of 'critical stages' of development. A leaf primordium which has passed the critical stage when pressurized may complete its growth, though only slowly, whereas a leaf which has not attained the critical point will suffer total arrest. Pressure treatment is not, however, totally unfavourable to growth. In treated plants the elongation of internodes already delimited is greatly increased; there is no significant increase in the adult size of cells, so presumably the treatment encourages internodal mitosis. Within the limits of Ferling's experiments the plants which survived pressurization appear to have made at least as much total growth as the unpressurized controls. Some of her plants, however, did not survive, possibly through mechanical damage arising from excessively abrupt pressure changes. The position is also complicated by seasonal differences in response.

In the stems of twining plants mechanical changes occur which are in other stems less prominent if not totally absent. The investigation of these aspects of growth is technically very difficult, calling not only for refined instrumentation but for considerable geometrical insight in the interpretation of results. One of the most critical studies was that of Hendricks (1919), in which the stem was gently stretched by a weight and pulley arrangement just enough to prevent any twining movement. Concurrent measurements were then taken of elongation, torsion, and torsional rigidity. This last was determined by making the stem the suspension of a torsion pendulum, the swinging bob of which was the mirror frame used in taking the angular measurement. These observations, continued through the whole early life of an internode in several species of twining plants, yield very consistently the pattern shown in Fig. 120 (see below). The early part of the record shows a steady increase in length accompanied by an equally progressive development of spiral twist, amounting eventually in some cases to more than three

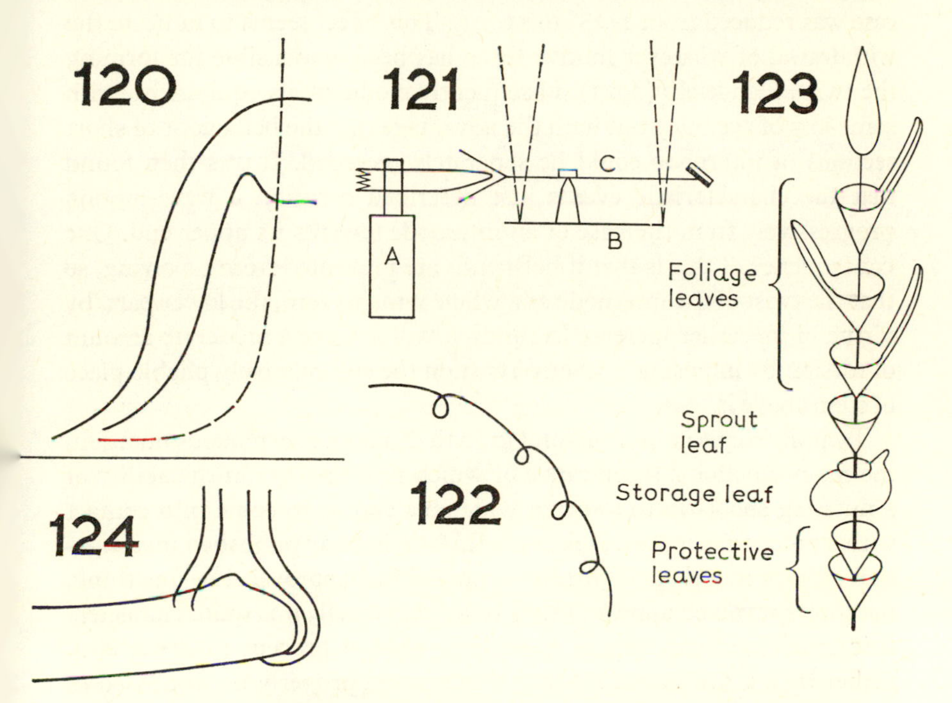

FIGURES 120–124. *Growth related to ecological requirements.*

120. Course of development in an internode of a twining stem, as indicated by observations of Hendricks (1919). Upper continuous graph represents sigmoid curve of growth in length, not given in original. Lower continuous curve is total twist of the internode, showing characteristic final relaxation. Broken line shows late but steep increase in torsional rigidity. **121.** Instrument used by Fisher to observe nutation in a rhizome apex. A is a small hydraulic jack with a clamping device. B is a platform suspended by threads (broken lines) from a support high above. B carries a fulcrum upon which rests the jewelled bearing of the lever C; this lever has at one end a conical socket to receive the thrust of the rhizome, at the other a mirror which forms the moving element of a high-magnification optical system. **122.** Type of trace obtained (not to scale); small loops on a much larger circle. **123.** Analysis of bulb by Mann. **124.** *Butomus* rhizome (simplified from Weber). The apex points vertically upwards, appearing as a slight prominence; growth of the rhizome is however entirely lateral, towards the right of the diagram; the bases of two leaves are shown.

complete revolutions of the upper end of the internode relative to its base. During the whole of this period there is only a very moderate rise in torsional rigidity. Then, with the cessation of growth in length, there is a sudden very large increase in rigidity, occasioned by the swift lignification of large masses of xylem. At about the same time there is a slight but curiously constant relaxation of the spiral twist (which in one typical case was reduced from 1128° to 1102°). This effect seems to indicate the withdrawal of whatever motive force has been responsible for forming the twist. Hendricks (1923) subsequently modified his equipment, with some loss of accuracy but with the advantage that the behaviour of short sections of internode could be separately recorded. It was then found that the characteristic events just described occur in a wave-motion progressively from the base of an internode towards its upper end. One consequence of this is that if both ends are prevented from revolving, so that the twist of the internode as a whole remains zero, the lower part, by virtue of its earlier increase in rigidity, will achieve a moderate amount of torsion by imposing a reverse twist on the still relatively pliable piece of stem above it.

Twining stems in their natural growth display a conspicuous nutation, the main functional significance of which probably lies in the ability of a nutating shoot-tip to sweep a wide area and so to come into contact with a potential support. It is doubtful whether any published investigation of such nutations is worthy of unqualified approval. For one thing, the forces involved appear always to be very small; it is quite characteristic that Hendricks should have been able to prevent nutation altogether by a gravitational loading which could properly be dismissed as negligible in relation to the torsion phenomena. There must accordingly be objections to the study of nutation by any method involving mechanical attachments to the plant. Many workers have therefore resorted to optical devices, but it then becomes extremely difficult to eliminate with certainty the effects of one-sided or intermittent illumination. Other problems arise from the growth of the plant in height, calling for some means whereby the relative levels of specimen and measuring instrument can be corrected at appropriate intervals (and with a due regard for the known sensitivity of young shoots to mechanical vibration). A promising system of instrumentation was described by Heathcote & Idle (1965). Two light levers were arranged to swing horizontally in the manner of gates, and spring-loaded so as to bear very gently upon the sides of a stem, the plant being so placed that the levers should cross always approximately at right angles. Each lever carries a mirror which reflects light from a fixed lamp to a recording-drum carrying photographic paper. With subsidiary equipment to print a time-base on the record, and with sufficient care to eliminate any non-linearities in the action of the lever system, a timed two-dimensional plot of the trajectory of the stem can be obtained by using the deflections of the light-spots directly as x- and y-coordinates. The plant material seems to have been fairly homogeneous genetically (*Phaseolus multiflorus* variety) and the conditions were standardized though somewhat unnatural (seedlings at 22°C. in low light-intensity). The measured tracks were nevertheless very variable. Some seedlings swung in smooth ellipses of fairly regular period, averaging about 163 minutes, but others yielded zigzag or entirely chaotic tracings. The authors concluded that there was no relationship between the direction of nutatory movement and any morphological feature of the seedling, but their assessment was perhaps less critical here than in some other matters; some of their published charts strongly suggest that maximum swing occurs in the plane defined by the midribs of the first pair of plumular leaves. Of much greater importance is the suggestion by Heathcote & Idle that the periodicity of nutation is more constant than the exact form of the orbit, and that periodicity may be ascertained by simpler means than those just outlined. When a stem is bent sideways, the result is a diminution in the height of the apex above the soil. Nutation in anything but a circular orbit thus introduces a cyclic fluctuation into the measured height of the plant. Such a fluctuation can very quickly and easily be detected by straightforward use of a cathetometer, and it is likely therefore that the laborious and uncertain plotting of orbital motions may for many purposes be dispensed with. Whether vertical measurements can be made to yield any information about the shape of the horizontal movement is not yet clear, but it is interesting that the speed of movement sometimes shows a systematic variation, as for instance in elliptical swings where orbital velocity is minimal at the ends of the major axis.

Another line of enquiry, ecologically very distinct, which also leads to a consideration of nutation, is the investigation of the penetration of soil by underground shoots. One of the most successful attempts was the study by Fisher (1964) of the rhizome of *Poa pratensis*. As in many other monocotyledons the rhizome here has a pointed tip consisting of specialized cataphylls which cover the apex of the stem. This type of rhizome tip appears well suited to a process of straight driving through the ground; that it is actually so driven does not of course follow. In *Poa* Fisher was able by means of ink-marks to establish that the rhizome was elongating only within a distance of 3 cm from the visible tip, and that it grew without any sustained rotation of the tip. Any gimlet action being thus excluded, it remained to estimate the power of penetration and to ascertain whether the rhizome tip exerts a straight thrust or whether, through an inherent tendency to circumnutate, it regularly changes the direction in which its apex bears upon the mineral framework of the soil. Fisher prepared standardized blocks of compressed clay. These were so hard that steel needles forced into them usually broke, yet the rhizomes grew fast and straight through the same obstacles. Moreover the first penetration of the hard surface of a clay block was not usually attended by any significant buckling of the subapical part of the rhizome, even though this, being in looser ground, might have been expected to show signs of a want of lateral support. These circumstances point strongly to the existence of some means whereby the tip is made more effective as a drill than an inert object of the same shape and strength could possibly be.

The principle of the instrument upon which Fisher relied for the rest of his enquiry is shown in Fig. 121. A is a refined hydraulic jacking device which clamps a rhizome at a point 3 cm from its tip and facilitates preliminary adjustment of this fixed position. B is a light platform which is suspended by four threads from a beam high above. The platform carries a vertical needle upon the point of which is balanced, by means of a jewelled watch-bearing, a light metal lever C, the two arms of which are respectively 3 and 24 mm. The shorter arm of C carries a conical socket which receives the thrust of the rhizome tip, while the longer arm bears a tiny mirror which reflects a spot of light on to a distant screen. The suspension of the platform B permits the whole lever assembly to ride forward upon the rhizome tip for as much as three days at a time, while the optical system is sufficiently sensitive to magnify the rhizome's deviations from a straight path by factors up to 10,000 as required. The path described by the tip of an exposed rhizome turns out to be a compound curve (Fig. 122), a loop about 2 μm in diameter being made about every forty minutes upon a circle of about 2 mm diameter which is completed in about twenty-four hours. A rhizome in firm soil cannot of course perform these movements, but there must be a very strong presumption that the direction of its thrust will change in a similar rhythm.

From the considerations so far presented in this chapter it appears doubtful whether the advance of a growing organ can ever be uniform or direct; in general we must be prepared to find a complex interaction involving circumnutatory movement, tropic stimulation and response, mechanical stresses arising from the plant's own weight or from the resistance of the environment, and the reactions of the tissues to those stresses. The entanglement of all these apparently separate phenomena is logically inescapable; it is not possible, for instance, to design any scheme of nutation which will not generate both a geotropic stimulus and a redistribution of bending-moments in the stem. It might, however, still be possible to argue that simpler explanations would suffice for resultant or average directions of growth: there would be no logical inconsistency in a contention that the upward growth of the leader shoot of a tree represented pure negative geotropism, into which nutation, wind action, and other more complex phenomena could introduce no more than transient and minor perturbations. Such views are widely held and in many cases may be very near to the truth. Nevertheless it has to be recognized that intensive examination of the causes underlying the general orientation of an organ has in a significant number of examples tended towards the destruction of the idea that the direction of growth expresses the dominance of one or two elementary tropisms.

Some plants display a significant orientation of organs, principally of the leaves, towards a particular point of the compass, and are accordingly known as compass-plants. Such effects appear always to be regulated by the incident illumination, but the mechanism involved is more complex than that of a simple phototropic response. Early observations on compass-plants were concerned with a very small number of species, particularly *Lactuca scariola* and a few other Compositae. These are tall plants in which the leaves are bent and twisted so that each lamina lies in a vertical, not a horizontal plane, and is moreover, in plants exposed to the sun, ordinarily placed in the plane of the meridian. The adaxial face of a leaf therefore does not face upwards but is turned either east or west, the two positions being about equally frequent. It is doubtful how widely this type of behaviour is distributed among other plants. Schanderl (1929), in describing a north–south orientation of the edges of the aerial leaves of *Sagittaria*, was clearly of the opinion that this might have escaped attention if the plants had been more crowded or the weather less sunny. Stocker (1926), in collating the experiences of travellers, thought that compass plant behaviour was rare in Mexico but rather common in South Africa. Later (Stocker, 1927) he modified this position recognizing that vertical orientation of the lamina, a common phenomenon in any hot and sunny climate, is by no means universally accompanied by a consistent horizontal orientation.

Early attempts to explain the behaviour of compass-plants were conceived in terms of reduction of transpiration and an avoidance of overheating of the leaf. The lamina, by turning its edge to the noon sun, was supposed to escape the various forms of physiological damage which excessive radiation might inflict. Huber (1935), working from this point of view, argued that the presentation of an edge to the sun would be relevant only to a broad flat leaf. If a leaf were so narrow that face and edge views were substantially similar in area the only remaining expedient for minimizing exposure would be for the leaf to point its tip at the noon sun. He gives a lively account of his examination of *Aster linosyris* in the Tyrol, which shows this effect, and which he called a 'gnomon-plant'. The orientation of the leaf in this case is such as to yield an estimate of the plant's geographical latitude. The idea that north–south orientation of the lamina will minimize the incidence of radiation does not appear to be well-founded. Dolk (1931) finds by actual calculation that in summer at all latitudes up to 60° a lamina with its edges pointing north–south receives much more light per day than one pointing east–west. Furthermore the intensity on an east–west plane at noon is no greater than the morning or evening maximum on a north–south plane. Admittedly these relationships are changed in winter, but that is hardly relevant to the botanical problem.

Dolk analysed the mechanism of response in *Lactuca*. Leaves emerging from the stem in an east or west direction are simply curved upwards, which brings them into the required meridian plane. Leaves emerging in the north or south directions have however to twist at the base. Twists appear to be autonomic, because they occur in every leaf of plants grown on a klinostat. Curvature of a leaf was considered by Dolk to inhibit the autonomic torsion. On the basis of his astronomical calculations, and supposing the plant to take up the posture which gives the greatest total interception of light, he grew a plant which received supplementary illumination from a heliostat on its south side. As expected, the result was an east–west orientation of the edges of the leaves, and a similar effect was obtained when plants were grown in a glasshouse in winter, which also implies a maximum radiation intake. There are records of east–west orientation in plants under natural conditions;

Stocker (1926) observed this in species of *Erodium* in Egyptian deserts in latitude 30°N. These leaves were somewhat unusual for the genus in having palisade tissue on both sides, and it seems to be rather generally the case that leaves which show distinct geographical orientation tend towards an equalization of anatomical construction on the two faces. The anatomical consequences of foliar orientation were more fully considered by Filzer (1942). In some previous work on tree crowns he had observed that leaves on the east side tend on the whole to resemble those on the north in possessing a 'shade leaf' construction, while those on the west side are more like those on the south in displaying 'sun leaf' characteristics. This state of affairs, in central Europe, cannot be due to a greater incidence of light on the west than on the east side of a tree. In fact, owing to the greater average cloudiness of the afternoon hours, the west side receives substantially less light than the east side. Filzer therefore attributes the anatomical effect to the unfavourable situation of the western leaves in relation to turgor. They receive their maximum illumination at a time when the temperature is high and the water-content of the tree has been depleted by many hours of day-time transpiration. The late afternoon sunshine therefore has a greater effect upon them even though it is not remarkable in absolute amount. In extending this principle to compass-plants, Filzer determined the distribution of stomata, which may be expected to occur less abundantly where the effective exposure to radiation is greater. In *Lactuca scariola* there are three leaf-positions to consider. In a horizontal leaf the stomata occur on upper and lower surfaces in the ratio of 70:130. In a leaf which stands vertically in the meridian with its 'upper' surface facing east the ratio is 100:100, whereas in a vertical leaf with its 'upper' surface facing west it is 94:106. This result is consistent with the supposition that a western exposure is more severe in its effects than its eastern counterpart.

It was demonstrated by the experiments of Schanderl (1932) that it is insufficient, in studying the orientation of leaves, to take into account only the direct illumination from sun and sky. Consideration must also be given to reflections from surrounding objects. Besides making many observations in natural habitats or in places where plants grew against a wall, Schanderl erected various structures which simulated conditions occurring in nature. Four-sided and eight-sided pyramids of slate were built, and plants were grown against the various faces. Some of the slate surfaces were blackened to reduce reflection, and use was also made of a sloping wall of glazed white tiles. As a general rule compass-plant effects are not prominent over surfaces which have poor reflecting properties, but a reflective horizontal substratum is favourable to north–south orientation. We may expect to find clear north–south arrangement of leaves over a water surface or level sand, but not usually over turf. Where the reflecting surface is placed asymmetrically around the shoot, including not only the proximity of such objects as a wall but also any pronounced slope of the ground, an orientation tends to develop in relation to the surroundings. Thus a plant on a west-facing steep slope of light-coloured bare soil will be likely to have an east–west orientation of its leaves. An orientation due to slope of the soil naturally tends to disappear at higher levels if the plant grows tall enough. In some of Schanderl's trials *Lactuca* plants showed east–west orientation up to a height of 50 cm, but leaves above that level were placed north–south. *Iris* rhizomes, on being transplanted between sloping sites of different geographical aspect, made a slow but decided adjustment of their leaf-orientation in the expected manner. As a result of his experiences, Schanderl, after a careful review of the theoretical ideas of his predecessors, rejected all of them, and came, probably correctly, to the belief that no simple physiological hypothesis can account for these effects.

Complex systems for controlling the direction of growth are also encountered in underground shoot systems, where the question of depth-regulation assumes considerable importance. The ecological requirements which the plant has to satisfy are not simple; it is not sufficient, for instance, that a rhizome should be endowed with such a response to gravity as would cause it to grow horizontally. Such a crude adaptation would prove insufficient in any case where the direct horizontal route was obstructed by a stone, or where the ground was sloping. The ability to respond to a variety of environmental factors in varying combinations would appear to be essential, and we cannot even assume at the outset that the underground shoot is entirely in control of its own actions; it has been suggested, for instance, that some species regulate the depth of their rhizomes by 'measuring up' the aerial shoot to see what length of the vertical stem is in darkness. The experimental investigation of these matters is attended with considerable technical difficulties. Thus Bennett-Clark & Ball (1951) at the commencement of their study of the rhizome of *Aegopodium*, were confronted with the fact that their material was too sensitive to light to permit any system of continuing visual observation. A thirty-second flash, even of dim red light, produced after a two-hour delay strong downward curvature of the rhizome, recovery from which would require at least another twenty hours. They were therefore obliged to record the behaviour of their specimens by an automatic camera, working by infra-red lamps in total visual darkness. As it was impracticable to excavate a rhizome and place it in the apparatus without exposure to light, twenty-four hours had to pass before any experiment could be started. The extreme sensitivity of these rhizomes to light does not constitute a phototropism because the direction of illumination is irrelevant to the direction of curvature, which is gravitationally determined. Nor is it likely that the reaction to light, interesting though it may be from the physiological point of view, can play any part in the normal life of the rhizome. For reasons which the Bennett-Clark & Ball investigation seems not to have revealed, these shoots ordinarily maintain a depth at which exposure to light can only occur through some exceptional accident; no doubt the response may in such an emergency possess some adaptive value. The observation that an *Aegopodium* rhizome turns upwards when exposed to an atmosphere enriched with carbon dioxide suggests a means by which rhizomes in nature may be prevented from going too deep. Rather surprisingly these rhizomes turn out to have no deeply engrained dorsiventral property. Turning a shoot upside-down causes oscillations which are spectacular but quite transitory, and there is no indication that the rhizome suffers any lasting inconvenience through being 'wrong way up'.

Highly standardized forms of shoot construction and behaviour are to be found in the more specialized types of perennating underground systems, such as bulbs. Mann (1960) has described the bulbs of *Allium neapolitanum* and some related species, in which the main axis of a year's growth has the arrangement shown in Fig. 123. The bulb consists of two 'protective leaves', which in this section of the genus become fused together by their adjoining epidermal faces, and a single 'storage leaf' which is so swollen as to occupy most of the volume of the bulb. The aerial part of the system arises by resumption of growth in the main apex of this bulb; this continuation of the stem carries a 'sprout leaf', which is a bladeless sheath barely reaching the surface of the soil, three foliage leaves, and a terminal inflorescence. Vegetative survival of the individual necessarily depends on the activity of axillary buds, one or more of which will normally grow into new bulbs. The situation is complicated, however, by the existence of a sharp distinction between two different types of lateral bulbs. The 'renewal' bulb, of which there is only one, in the axil of the uppermost foliage leaf, reproduces exactly the structure indicated in Fig. 123, and precisely replaces the original bulb. In addition there may be one or more 'increase' bulbs, formed in the axils of lower leaves. Increase bulbs are much smaller than the renewal bulb, with a reduced quota of foliage, and are further set apart by the fact that each possesses a prophyll, a scale leaf preliminary and external to the two fused protective leaves; the renewal bulb unaccountably lacks this feature. Holttum (1955) has argued with considerable plausibility that highly condensed sympodial shoot systems such as that of *Allium* may owe their evolutionary origin in part to the anatomical peculiarities of the monocotyledonous stem. His idea is that the lack of cambium and the consequent inability to increase beyond a certain limit the amount of vascularization in a given transverse section must impose some restraint on the elongation of internodes. Upon this view the characteristic monocotyledonous habit, with a more or less tuberous stem and a heavy dependence on adventitious rooting, has to some extent been forced upon the plant. This leads to the interesting speculation that types of sympodial branching which were particularly suited to the problems of monocotyledons with continuous growth in moist tropical climates may have constituted a first stage in the evolution of bulbs and other forms adapted to dry-season survival.

Some underground organs are morphologically indeterminate, as for example the tubers of *Dioscorea* which were examined by Martin & Ortiz (1963). In the seedling the tuber begins as a lateral swelling of the hypocotyl and in the normal course all subsequent tuber growth arises by lobing of the original tuber, though new tubers can exceptionally be obtained from stem or leaf-cuttings. The tuber is lacking in morphological differentiation; it has no leaves and branches irregularly, sometimes forming an approximate dichotomy but never producing any of the characteristic features of a node. The tuber is, however, in some species the only really permanent part of the plant. Martin & Ortiz were inclined to regard the tuber, which has rather more anatomical resemblance to a stem than to a root, as a greatly modified rhizome. Such an interpretation appears somewhat strained, particularly as there is a sharp taxonomic distinction between *Dioscorea* species which have tubers and others which have normal rhizomes. Mixed or intermediate conditions have not yet been found.

Just as in aerial shoots, the extreme monocotyledonous condition, in which short internodes are associated with great primary thickening of the stem, can involve complex problems of growth regulation. Weber

(1950) has described the rhizome of *Butomus*, the apical region of which is shown in Fig. 124. The growth of the rhizome is horizontal, but its apical dome faces directly upwards. As this system is monopodial, with its main apex growing onwards year after year without ever emerging from the soil, we have to reckon with a permanent right-angle bend of the axis. Penetration of the soil is not the business of the apex at all but is carried out by a lateral flank. The apex is not in the place where external inspection of a specimen would lead one to seek it, and the longitudinal growth of the stem is so thoroughly confounded with its peculiar processes of thickening that the shoot makes absolutely no advance in the direction in which its tip is pointed.

In some of the specialized subterranean shoot system in which the growth of the individual stem axis is predominantly vertical (mostly bulbs and corms) an important part is played by contractile roots, the function of which in most cases is merely to drag the shoot system downwards by an amount which will compensate for upward growth, so preventing the plant from growing out of the soil. Although the behaviour of the root itself is of great interest, the displacement of the shoot is in most of these cases too small to produce any spectacular effect. An outstanding exception can be seen in *Oxalis cernua*, described by Galil (1968). The organ of perennation is here a small elongated bulb, from which a shoot grows vertically upwards to the surface of the soil, and there produces a crown of leaves. The peculiarities of vegetative propagation in this species are indicated in Fig. 125 (see p. 210), where the old bulb, and the crown produced from it, are shown at the right. With the withdrawal of reserve materials from the bulb, it disintegrates in the usual way; its inner (storage) cataphylls collapse, and although its outer (protective) cataphylls may endure for some time their connections with the basal stem-disc of the bulb become very fragile. The lower part of the stem which leads up to the foliage rosette is endowed with a great capacity for elongation, and unless the original planting was abnormally deep there comes a time when a surplus length of this thin stem (Galil calls it 'the thread') may be crumpled within the shrivelled husk of the old bulb. From the base of the bulb there grows out a single fat root, which extends almost horizontally, creating a tunnel about 7 mm in diameter and up to 50 cm long (though 30 cm would be more usual). This root is contractile in the most exceptional degree, shrinking often to no more than 30% of its original length. The thread-like stem is thereby dragged through the bottom of the old bulb and down the tunnel. As

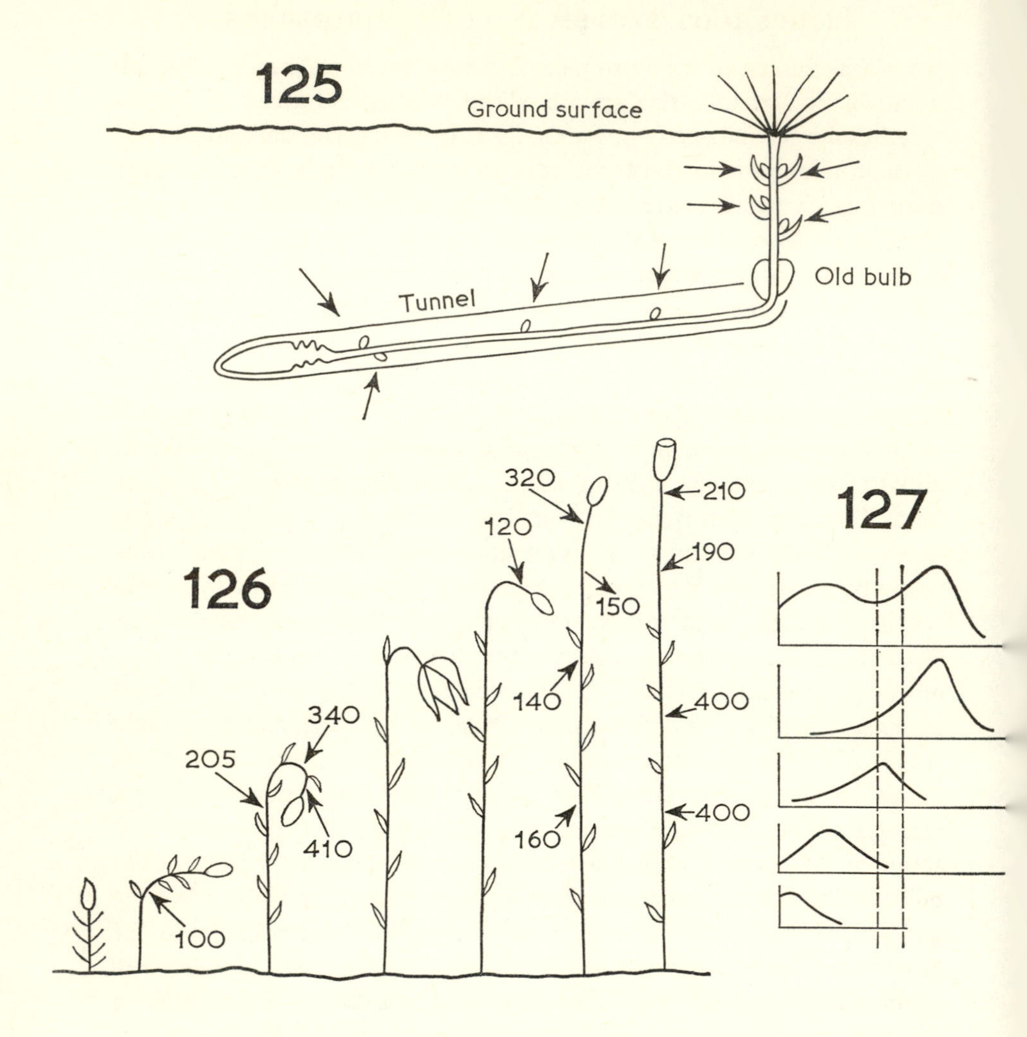

FIGURES 125–127. *Specialized patterns of behaviour.*

125. Subterranean organs of *Oxalis cernua*, observed by Galil. The old bulb has given rise to a rosette at ground level. The vertical stem leading up to the rosette has grown to excessive length. For a time it was coiled in the husk of the old bulb, but then a contractile root grew out and created the tunnel; when the root contracted the excess length of stem was pulled down the tunnel. New bulbs are now arising at the points marked with arrows. **126.** Successive stages in the growth of the flowering shoot of *Fritillaria*, observed by Kaldewey. Attached figures are auxin concentrations. **127.** Curves of growth-rate against time, the five-day period of anthesis marked by vertical broken lines. The four lower single-peaked curves refer to individual zones of stem from the base upwards. The double-peaked curve is the total growth, to which the uppermost internode makes a disproportionate contribution.

new bulbs arise at such points as are indicated by arrows in Fig. 125, this constitutes an effective mechanism for short-range dispersal.

Other shoots liable to display complicated patterns of behaviour are those concerned with the movement of reproductive organs into their proper positions. We may take as an example the flowering shoot of *Fritillaria meleagris*, which was thoroughly investigated by Kaldewey (1957). The shoot here rises from a bulb, and bears a few leaves alternately, followed by a single terminal flower. As the shoot emerges from the soil the flower-bud is erect (Fig. 126), but at a very early stage the stem assumes a curvature at the level of its lowest leaf and the bud is thereby turned downwards through an angle which ultimately exceeds 180°. The bend in the stem persists until the flower has been pollinated and the development of the fruit is well advanced, but it does not stay in the same place. At the time of flowering the bend is at the level of the uppermost leaf, so that several leaves pass over the crest of the bend in the course of prefloral growth. After fertilization the bend straightens out, and there is rapid elongation of the internode above the last foliage leaf.

The total height growth of this shoot displays two prominent maxima, with the five-day period of anthesis falling in the trough between them (Fig. 127). There is not, however, a double peak of longitudinal growth at any one level in the stem. Successively higher internodes are later in reaching their individual growth maxima, and the second crest of the resultant curve arises only because the activity of the uppermost internode is disproportionate. Kaldewey, on the grounds of observations on the subepidermal layer of cells, claims that this post-floral surge of growth in the top internode is entirely a matter of cell-enlargement, an elongation of cells by a factor of about 10. Removal of the ovary, or of the whole flower bud, or even a failure to effect fertilization, will prevent the normal post-floral elongation, and the deficiency can be made good by artificially supplying growth hormones. This suggests that the ovary in its later development may be a source of auxin. The distribution of growth-hormone activity is shown in Fig. 126, where a low base-level of 80 units or less must be understood to prevail in the absence of other figures. These results were obtained by a technique of biological assay, and they are generally concordant with the natural distribution of growth. There are two periods during which substantial amounts of growth hormones are flowing in the stem, and these correspond with the two elongation maxima. Upon more detailed examination some finer points emerge. At the right of Fig. 126 it can be seen that very high concentrations of hormone build up in the middle region of the stem, but no elongation takes place here, the tissues no longer being sufficiently immature to respond. The plant here produces an excess of hormone to no apparent purpose. By contrast, the cells in the lowest internodes do not in intact plants attain the lengths to which they can be forced to grow by artificially supplying auxin to decapitated stems. In the early stages, therefore, it appears that the plant's performance is limited by an inherent deficiency of auxin. A particularly interesting feature is the production by the stamens in the period before the opening of the flower of growth-inhibitors which partly offset the effects of auxin coming from other parts of the flower, and so contribute to the arrest of the prefloral stage of elongation.

The curvature of the stem can easily be attributed to differences in the elongation of cells at a fairly early stage in their growth. Kaldewey found that cells on the convex side of the bend were usually about 40% longer than those on the concave side. This accounts for the observed curvature, but most of the elongation of the cells is still to come. To carry the explanation further is more difficult. Putting plants on a horizontal klinostat makes the stems grow straight, but it does not render the whole mechanism inoperative, for a plant removed from the klinostat very quickly takes up the curvature appropriate to its stage of development, indicating that some of the characteristic changes continue even though their effects are concealed. As the mature stem shows no radial inequality of cell size the straightening of each piece as its growth comes to an end may be regarded as more or less automatic. Curvature will arise, without any difference in the total growth to be accomplished, if the cells on the convex side become responsive to auxin earlier than those on the concave side. On this basis, as the curvature is only to be temporary, it is not necessary that there should be any asymmetry in the radial distribution of auxin. In *Fritillaria* any treatment (e.g. removal of flower, removal of ovary, prevention of fertilization) which reduces the auxin content of the plant causes premature straightening of the stem which can be partly delayed by artificial supply of auxin.

Some species have long been noted as extreme examples of aerial shoot dimorphism. Thus in *Hedera helix* there is a sharp distinction between the prostrate or climbing shoots, with distichous phyllotaxy and lobed leaves, and the erect type in which spiral phyllotaxy prevails and the leaves have a simpler outline without lobes. The production of flowers is exclusively the prerogative of the erect shoots, and the two forms are in fact distinguishable by a number of histological characters as well as the obvious points of phyllotaxy and leaf-shape. It was long ago established that the two shoot forms display a marked persistence in vegetative propagation; by normal horticultural practice each can be maintained by the repeated rooting of stem-cuttings. The constancy of the two forms is not absolute; each will occasionally give rise to a shoot of the other, but the degree of stability which is observed is sufficient to pose a considerable physiological problem. In the natural life of the species the distichous shoot appears as a juvenile form, the spiral shoot as the adult. Experience with garden plants shows that the juvenile form, whether established as a cutting or from seed, must usually grow for ten or twelve years before it can produce an adult branch. The adult shoot shows a noteworthy diminution in the remarkable facility for producing adventitious roots which otherwise distinguishes this species (Frank & Renner, 1956, in preparing material for their physiological experiments, were never able to push their success rate for rooting adult-form cuttings above 14%). Physiological interest has largely centred on the discovery of treatments which would cause the adult form to revert to the juvenile state. Such reversion is fairly uncommon, both in the wild and in normal cultivation. A number of investigators have claimed to be able to increase the frequency of reversion by various processes; in fact none of the recommended treatments appears to be completely reliable (in the sense that all further growth is of juvenile type). Many of them, however, have a significant effect.

Frank & Renner obtained no success upon treating adult shoots with IAA or with hydrocyanic acid, but were able to cause reversion by subjecting plants to X-ray treatment or to low-temperature shock. They also conducted experiments in which an adult shoot shared a water-culture vessel with a juvenile one. In these conditions the adult shoot often reverts to the juvenile state, but there is no reciprocal effect; the juvenile shoot is not induced to grow into the mature form. This evidence clearly points to the existence of a water-soluble 'juvenile hormone'.

The action of such a hormone would also account for the effects reported by Stoutemeyer & Britt (1961) from a series of grafting experiments. Adult shoots were both grafted and budded on juvenile stock, the distinction lying in the amount of adult tissue transferred, which is much less in budding than in grafting operations. Two temperatures of incubation were used, and the percentages of the adult samples which reverted to the juvenile state in subsequent growth were: grafted plants 3% at 60°F. but 25% at 80°F.; budded plants 15% at 60°F. but 97% at 80°F. Evidently the influence of the juvenile stock can more readily overthrow the 'adult' equilibrium of the small bud scion than that of the larger graft scion, in which presumably there is space for any hormone conveyed from the stock to become diluted or depleted before it can reach the meristems. To this is added a significant temperature effect.

It seems not to be known whether the juvenile forms of *Hedera* which are obtained by naturally occurring or experimentally induced reversion from the adult state would display the same measure of stability as juvenile shoots obtained from seed. It is interesting, however, that Robbins (1957, 1960), who obtained reversion by the application of gibberellic acid, took tip-cuttings from his reverted shoots, grew these on without further exposure to the acid, and found that they persisted in the juvenile form long after it had ceased to be credible that any of the applied acid could still linger in the specimens. This tends to confirm the impression that the shoot in these plants has two stability states, and that a specimen once pushed into either of these tends to persist in it. The distinction, although clear-cut in most *Hedera* stocks, is obscured in some specimens by the occurrence of intermediate shoot forms showing various degrees of permanence. In so far as these aberrations have been investigated, they appear to have a genetical basis. Thus Robbins (1960) reported on a variety '238th Street', in which the juvenile form, although distinguished in the normal way by distichous phyllotaxy and by abundant rooting, carries inflorescences and lacks the normal lobing of the leaf.

Shoot dimorphism of a different category is widespread among trees and shrubs, and appears to be related to the geometrical considerations which arise in producing a large foliage canopy from small leaves. Given the proportions which exist in the most actively elongating shoots of many woody plants between leaf-area and internode length, and remembering that a woody stem of any age must be leafless for most of its length, it is perfectly obvious that an adequate photosynthetic system could not be built up from this type of shoot alone. The requisite leaf-area can only be obtained by the production of leaves on lateral branches in which there will have to be a restraint on internodal elongation. To put the matter in another way, the greater the emphasis on elongation of the main stem, the more pronounced will be the effects of apical dominance, and the more likely are the lateral shoots to display conspicuous morphological peculiarities. In a large number of cases the long shoots and the short shoots constitute two separate classes. A good example of this rather extreme division of function is seen in the work of Titman & Wetmore (1955) on *Cercidiphyllum*. Here the young tree produces predominantly long shoots, but at the age of 15–20 years flowering begins, and the long shoot structure is thereafter to be found only in the leader shoot of each major division of the shoot system. Long shoots are monopodial, and have opposite leaves. Short shoots are always sympodial and although about a dozen foliar primordia are formed in the season only one of these develops into a foliage leaf. The rest of the growth increment of that year either forms an inflorescence, or, if vegetative, it merely aborts. Stem elongation in short shoots is minimal, and even ten years of growth produces only a small stem spur. There are considerable differences of shape, as is common in these cases, between the leaves of long and short shoots respectively. Although the structural differences are impressive, this kind of shoot dimorphism seems to have a much simpler kind of physiological basis than the semi-permanent conditioning which determines the two forms of shoot in *Hedera*. The short shoot/long shoot contrast appears to be directly controlled by auxin balance and to represent merely an exaggerated form of apical dominance. With that dominance removed, a short shoot of *Cercidiphyllum* will go over completely to long-shoot behaviour in a single season.

Situations which are similar in principle though very different in appearance can be found in monopodial rhizomes. Webster & Steeves (1958) have described a remarkable state of affairs in *Pteridium aquilinum* where the long shoots of the rhizome system appear to be genuinely leafless (a very rare condition in living vascular plants), the apex being protected only by a coating of hairs. The lateral short shoots produce leaves (normally at the rate of one a year) separated by internodes up to 2 cm long, but still with the tip of the shoot projecting some way beyond the youngest recognizable leaf. In the long shoot it is meaningless to think of internode length, but the annual growth of the long shoot, which is no more than 17 cm in the American variety *latiusculum*, can in the variety *aquilinum* growing in Scotland be as much as 90 cm, with important ecological consequences. The relatively exposed position of the apices in this system, though perhaps not very rare in ferns, is unlikely to have many close counterparts among the angiosperms.

In studying the general form of branched shoot systems it may be important to distinguish between cases with winter dormancy and those in which only a single season's growth has to be considered. In herbaceous systems there is sometimes a very simple and direct relationship between branching and the production of hormones. Delisle (1937) made comparisons between *Aster novae-angliae*, which branches little, *A. multiflorus*, which branches much more, and the hybrid, which is intermediate. Taking the auxin-output of the stem tip or 'terminal bud' in *A. novae-angliae* as 100, the comparative value for *A. multiflorus* is 74 and that for the hybrid is 84. It is necessary that the assay procedure should be applied strictly to the part of the shoot which is too young to contain any lateral branches which are themselves active as sources of auxin. If too large a specimen be taken the order of the species, in respect of auxin production, is reversed. From this point of view Delisle's definition of the 'bud' (really only the unexpanded apical part of the shoot) is not altogether satisfactory. That equally simple relationships can for long prevail in the crown of a tree is much more doubtful. The form of the branch-system in a tree is determined primarily by the selection of buds which are to grow into elongated shoots, as distinct from those which will remain dormant or merely form short spurs. The excurrent tree form, which has a single well-marked leader shoot, and which is characteristic of many coniferous species, is rather sharply separable from the decurrent type, in which there is no leader but only a large number of twigs of roughly equal status. Passage from one condition to the other in the course of life is common; indeed a very young tree can can hardly be other than excurrent. It is not at all easy, however, to secure by pruning any lasting command of the habit of growth. By pruning we may very well be able to control the short-term future of particular individual branches, but the general statistical balance of branching behaviour is unaffected by this consideration, and unless the development of the system is under constant supervision and correction the species will quickly revert to its characteristic habit. A tree with excurrent propensities which is deprived of its leader will soon establish another, whereas in a truly decurrent type an artificially created leader is liable to lose its primacy as soon as pruning is discontinued. These differences have often been attributed to a greater intensity of apical dominance in excurrent than in decurrent trees. It appears, however, that this view may be oversimplified if not totally incorrect, and it is certainly unsafe to extend to the larger problems of tree growth any principle of apical dominance which has only been demonstrated in the more localized context of long shoots and spurs. It has been pointed out (Brown, McAlpine & Kormanik, 1967) that when observation is restricted to the year in which the main axis elongates, the development of lateral branches is paradoxically greater in excurrent than in decurrent types. A possible form of hypothesis would be that the lateral buds on a twig of a decurrent tree, being held completely dormant in the year of their formation, begin the next season on a fairly equal footing with the (also dormant) terminal bud, whereas the laterals of an excurrent leader, having been permitted in their first season to make limited growth in a strictly subordinate capacity, have acquired a semi-permanent conditioning which prevents them from catching up with an undisturbed leader.

An important part of the study of environmental influences is concerned with the existence of different response mechanisms in different organs at the same time, and many relevant examples can be found in modern physiological literature. Thus it was found by Leshem & Koller (1965) that in strawberry plants both the number and length of the runners are increased by raising the temperature of the roots, but that there was no corresponding advantage in warming the rest of the plant. A comparable form of organ-specificity can be seen in the examination by Sparmann (1961) of plants grown from vernalized and unvernalized seeds of barley. In Table 18 we have the lengths at maturity, in mm, of lamina and sheath of the first eight leaves of the main shoot.

Calculating in a manner not adopted in the original, we obtain the effect of vernalization as a percentage increase in length of sheath or lamina at each node. It can be seen that the two parts of the leaf respond in quite different ways. The analysis can be carried a stage further, as shown in the last two lines of the table. Here it appears that the general trend is for vernalization to produce a moderate reduction in the length of the organs at the base of the shoot, but a more pronounced increase at higher levels. Superimposed upon this pattern is a divergence between sheath and lamina which becomes progressively more important as growth proceeds.

Not only may particular forms of response be quite sharply localized within the body of the plant, but the responses may be linked with quite specialized aspects of the environment: for example Leshem & Koller were able to demonstrate a marked increase in length (though not in number) of strawberry runners subjected to strongly fluctuating temperatures with a cold night period. Whatever skill may be brought to bear, the discovery of such relationships must always be to some extent

TABLE 18

Effects of vernalization on leaves of barley seedlings, from observations by Sparmann (1965). Lengths of mature lamina and sheath for first eight leaves in millimetres

	Number of leaf							
	1	2	3	4	5	6	7	8
Lamina vernalized	41	90	154	205	244	282	270	180
control	70	140	176	230	250	265	250	256
Percentage increase (A)	−41	−36	−12	−11	−2	+6	+8	−30
Sheath vernalized	27	44	70	100	123	148	160	230
control	32	40	45	70	84	79	81	83
Percentage increase (B)	−16	+10	+56	+43	+46	+87	+98	+177
Average effect of vernalization: ½(A+B)	−29	−13	+22	+16	+22	+41	+53	+74
Emphasis on sheath: (B−A)	25	46	68	54	48	81	90	147

fortuitous. There is always a danger that physiological mechanisms which are influential in nature may be inactivated, as in this case, by some apparently innocent precaution such as the application of a constant temperature, or that experimental treatments may give negative results because they are applied to the wrong part of the plant, or not quite at the right time. It is hardly possible at present to bring into any coherent scheme the morphological aspects of such studies, although the number of relevant observations is large and increasing.

References

Abbe, E. C., Randolph, L. F. & Einset, J. (1941). The developmental relationship between shoot apex and growth pattern of leaf blade in diploid maize. *Am. J. Bot.* **28**, 778–784.

Allsopp, A. (1955). Experimental and analytical studies of pteridophytes. XXVII. Investigations on *Marsilea*. 5. Cultural conditions and morphogenesis, with special reference to the origin of land and water forms. *Ann. Bot.* N.S. **19**, 247–264.

Amer, F. A. & Williams, W. T. (1957). Leaf-area growth in *Pelargonium zonale*. *Ann. Bot.* N.S. **21**, 339–342.

Arney, S. E. (1955). Studies of growth and development in the genus *Fragaria*. IV. Winter growth. *Ann. Bot.* N.S. **19**, 265–276.

Arney, S. E. (1955a). Studies of growth and development in the genus *Fragaria*. V. Spring growth. *Ann. Bot.* N.S. **19**, 277–287.

Arnott, H. J. (1959). Anastomoses in the venation of *Ginkgo biloba*. *Am. J. Bot.* **46**, 405–411.

Arnott, H. J. & Tucker, S. C. (1963). Analysis of petal venation in *Ranunculus*. I. Anastomoses in *R. repens* var. *pleniflorus*. *Am. J. Bot.* **50**, 821–830.

Arnott, H. J. & Tucker, S. C. (1964). Analysis of petal venation in *Ranunculus*. II. Number and position of dichotomies in *R. repens* var. *pleniflorus*. *Bot. Gaz.* **125**, 13–26.

Ashby, E. (1948). Studies in the morphogenesis of leaves. I. An essay on leaf shape. *New Phytol.* **47**, 153–176.

Ashby, E. (1950). Studies in the morphogenesis of leaves. VI. Some effects of length of day upon leaf shape in *Ipomoea caerulea*. *New Phytol.* **49**, 375–387.

Bain, H. F. (1940). Origin of adventitious shoots in decapitated cranberry seedlings. *Bot. Gaz.* **101**, 872–880.

Balfour, E. E. & Philipson, W. R. (1962). The development of the primary vascular system of certain dicotyledons. *Phytomorphology* **12**, 110–143.

Bennett-Clark, T. A. & Ball, N. G. (1951). The diageotropic behaviour of rhizomes. *J. exp. Bot.* **2**, 169–203.

Berg, A. R. & Cutter, E. G. (1969). Leaf initiation rates and volume growth rates in the shoot apex of *Chrysanthemum*. *Am. J. Bot.* **56**, 153–159.

Bergdolt, E. (1934). Über die Bedingungen der Phyllodienbildung bei *Acacia* und über Licht-Reizbewegungen an *Oxalis rusciformis*-Phyllodien. *Flora, Jena.* **127**, 362–379.

Bierhorst, D. W. (1959). Symmetry in *Equisetum*. *Am. J. Bot.* **46**, 170–179.

Bindloss, E. A. (1942). A developmental analysis of cell length as related to stem length. *Am. J. Bot.* **29**, 179–188.

Blackman, V. H. (1919). The compound interest law and plant growth. *Ann. Bot.* **33**, 353–360.

Bostrack, J. M. & Millington, W. F. (1962). On the determination of leaf form in an aquatic heterophyllous species of *Ranunculus*. *Bull. Torrey bot. Club* **89**, 1–20.

Bowes, B. G. (1961). Inequality in the development of the axillary members in *Glechoma hederacea* L. *Ann. Bot.* N.S. **25**, 391–406.

Braun, M. (1957). Zur Kenntniss von *Epilobium* besonders *E. alpestre* (*trigonum*). I. Die Vererbung der Gliederzahl in den Blattwirteln. *Planta* **50**, 144–176.

Brett, D. W. & Dormer, K. J. (1960). Observations on a cyclic fluctuation in the leaf serrations of *Spiraea salicifolia*, and on the asymmetry of the leaf. *New Phytol.* **59**, 104–108.

Brotherton, W. B. & Bartlett, H. H. (1918). Cell measurement as an aid in the analysis of quantitative variation. *Am. J. Bot.* **5**, 192–206.

Brown, C. L., MacAlpine, R. G. & Kormanik, P. P. (1967). Apical dominance and form in woody plants: a reappraisal. *Am. J. Bot.* **54**, 153–162.

Brown, S. W. (1944). Studies of development in larkspur. I. Form sequence in the first ten mature leaves. *Bot. Gaz.* **106**, 103–108.

Carlquist, S. (1957). Leaf anatomy and ontogeny in *Argyroxiphium* and *Wilkesia* (Compositae). *Am. J. Bot.* **44**, 696–705.

Charlton, W. A. (1968). Studies in the Alismataceae. I. Developmental morphology of *Echinodorus tenellus*. *Can. J. Bot.* **46**, 1345–1360.

Church, A. H. (1904). *On the relation of phyllotaxis to mechanical laws.* 353 pp. (Williams & Norgate, London).

Critchfield, W. B. (1960). Leaf dimorphism in *Populus trichocarpa*. *Am. J. Bot.* **47**, 699–711.

Cunnell, G. (1961). The morphology of the inflorescence in *Ranunculus bulbosus* L. *Ann. Bot.* N.S. **25**, 224–240.

Cutter, E. G. (1962). Regeneration in *Zamioculcas*: an experimental study. *Ann. Bot.* N.S. **26**, 55–70.

Cutter, E. G. (1967). Morphogenesis and developmental potentialities of unequal buds. *Phytomorphology* **17**, 437–445.

Danert, S. (1953). Über die Symmetrieverhältnisse der Acanthaceen. *Flora, Jena* **140**, 307–325.

Davies, P. A. (1939). Leaf position in *Ailanthus altissima* in relation to the Fibonacci series. *Am. J. Bot.* **26**, 67–74.

Dede, R. A. (1962). Foliar venation patterns in the Rutaceae. *Am. J. Bot.* **49**, 490–497.

Delisle, A. L. (1937). The influence of auxin on secondary branching in two species of *Aster*. *Am. J. Bot.* **24**, 159–167.

Denne, M. P. (1966). Leaf development in *Trifolium* repens. *Bot. Gaz.* **127**, 202–210.

REFERENCES

Dolk, H. E. (1931). The movement of the leaves of the compass-plant *Lactuca scariola*. *Am. J. Bot.* **18**, 195–204.

Dore, J. (1955). Studies in the regeneration of horseradish. I. A re-examination of the morphology and anatomy of regeneration. *Ann. Bot.* N.S. **19**, 127–137.

Dormer, K. J. (1944). Some examples of correlation between stipules and lateral leaf traces. *New Phytol.* **43**, 151–153.

Dormer, K. J. (1945). An investigation of the taxonomic value of shoot structure in angiosperms with especial reference to Leguminosae. *Ann. Bot.* N.S. **9**, 141–153.

Dormer, K. J. (1945a). On the absence of a plumule in some leguminous seedlings. *New Phytol.* **44**, 25–28.

Dormer, K. J. (1946). Vegetative morphology as a guide to the classification of the Papilionatae. *New Phytol.* **45**, 145–161.

Dormer, K. J. (1950). Observations on the vascular supply to axillary branches *New Phytol.* **49**, 36–39.

Dormer, K. J. (1950a). A quantitative study of shoot development in *Vicia faba*. I. The xylem of the plumule. *Ann. Bot.* N.S. **14**, 421–434.

Dormer, K. J. (1951). A quantitative study of shoot development in *Vicia faba*. III. The dry weights of the plumular internodes. *Ann. Bot.* N.S. **15**, 289–303.

Dormer, K. J. (1954). Observations on the symmetry of the shoot in *Vicia faba* and some allied species, and on the transmission of some morphogenetic impulses. *Ann. Bot.* N.S. **18**, 55–70.

Dormer, K. J. (1954a). The acacian type of vascular system and some of its derivatives. I. Introduction, Menispermaceae and Lardizabalaceae, Berberidaceae. *New Phytol.* **53**, 301–311.

Dormer, K. J. (1955). *Asarum europaeum* – a critical case in vascular morphology. *New Phytol.* **54**, 338–342.

Dormer, K. J. (1965). Correlations in plant development: general and basic aspects, pp. 452–478 in Ruhland, W. *Encyclopaedia of Plant Physiology* 15 (1). 1647 pp. (Springer, Berlin).

Dormer, K. J. & Bentley, J. A. (1952). Some complex relationships between auxin content and leaf area in *Ipomoea caerulea* Koen. *New Phytol.* **51**, 116–126.

Dormer, K. J. & Cutter, E. G. (1959). On the arrangement of flowers on the rhizomes of some Nymphaeaceae. *New Phytol.* **58**, 176–181.

Dormer, K. J. & Hucker, J. (1957). Observations on the occurrence of prickles on the leaves of *Ilex aquifolium*. *Ann. Bot.* N.S. **21**, 385–398.

Dormer, K. J. & Plack, R. L. (1951). A quantitative study of shoot development in *Vicia faba*. II. The elongation of internode and petiole. *Ann. Bot.* N.S. **15**, 157–173.

Dostál, R. (1931). Versuche über die Massenproportionalität bei der Regeneration von *Bryophyllum crenatum*. *Flora, Jena* **124**, 240–300.

Erickson, R. O. (1966). Relative elemental growth rates and anisotropy of growth in area: a computer programme. *J. exp. Bot.* **17**, 390–403.

Erickson, R. O. & Michelini, F. J. (1957). The plastochron index. *Am. J. Bot.* **44**, 297–305.

Ertl, P. O. (1932). Vergleichende Untersuchungen über die Entwicklung der Blattnervature der Araceen. *Flora, Jena* **126**, 115–248.

Evans, A. T. (1928). Vascularization of the node in *Zea mays*. *Bot. Gaz.* **85**, 97–103.

Evans, P. S. (1965). Intercalary growth in the aerial shoot of *Eleocharis acuta* R. Br. Prodr. I. Structure of the growing zone. *Ann. Bot.* N.S. **29**, 205–217.

Ezelarab, G. E. & Dormer, K. J. (1963). The organization of the primary vascular system in Ranunculaceae. *Ann. Bot.* N.S. **27**, 23–38.

Ezelarab, G. E. & Dormer, K. J. (1966). The organization of the primary vascular system in the Rhoeadales. *Ann. Bot.* N.S. **30**, 123–132.

Fahn, A. & Arzee, T. (1959). Vascularization of articulated Chenopodiaceae and the nature of their fleshy cortex. *Am. J. Bot.* **46**, 330–338.

Fahn, A. & Broido, S. (1963). The primary vascularization of the stems and leaves of the genera *Salsola* and *Suaeda* (Chenopodiaceae). *Phytomorphology* **13**, 156–165.

Ferling, E. (1957). Die Wirkung des erhöhten hydrostatischen Druckes auf Wachstum und Differenzierung submerser Blütenpflanzen. *Planta* **49**, 235–270.

Filzer, P. (1942). Beiträge zum Kompasspflanzenproblem und einigen verwandten Problemen. *Flora, Jena* **135**, 435–444.

Fisher, J. E. (1964). Evidence of circumnutational growth movements of rhizomes of *Poa pratensis* L. that aid in soil penetration. *Can. J. Bot.* **42**, 293–299.

Fosket, E. B. (1968). The relation of age and bud break to the determination of phyllotaxy in *Catalpa speciosa*. *Am. J. Bot.* **55**, 894–899.

Foster, A. S. (1950). Morphology and venation of the leaf in *Quiina acutangula* Ducke. *Am. J. Bot.* **37**, 159–171.

Frank, H. & Renner, O. (1956). Über Verjüngung bei *Hedera helix* L. *Planta* **47**, 105–114.

Fron, M. G. (1899). Recherches anatomiques sur la racine et la tige des Chénopodiacées. *Annls. Sci. nat.* s. 8 Bot. **9**, 157–240.

Galil, J. (1968). Vegetative dispersal in *Oxalis cernua*. *Am. J. Bot.* **55**, 68–73.

Gertrude, M.-T. (1937). Action du milieu extérieur sur le métabolisme végétal. VIII. Métabolisme et morphogénèse en milieu aquatique. *Revue gén. Bot.* **49**, 161–181, 242–298, 328–352, 375–400, 449–467.

Gessner, F. (1940). Beiträge zur Biologie amphibischer Pflanzen. *Ber. dt. bot. Ges.* **58**, 2–22.

Girolami, G. (1953). Relation between phyllotaxis and primary vascular organization in *Linum*. *Am. J. Bot.* **40**, 618–625.

Gompertz, B. (1825). On the nature of the function expressive of the law of human mortality, and on a new method of determining the value of life contingencies. *Phil. Trans. roy. Soc.* B **115**, 513–585.

Goodwin, R. H. & Stepka, W. (1945). Growth and differentiation in the root tip of *Phleum pratense*. *Am. J. Bot.* **32**, 36–46.

Götz, O. (1953). Über die Brutknospenentwicklung der Gattung *Bryophyllum* im Langtag und Kurztag. *Z. Bot.* **41**, 445–482.

Gray, J. (1928). The kinetics of growth. *Brit. J. exp. Biol.* **6**, 248–274.

Gregory, F. G. (1921). Studies in the energy relations of plants. I. The increase in area of leaves and leaf surface of *Cucumis sativus*. *Ann. Bot.* **35**, 93–123.

Gregory, F. G. (1928). Studies in the energy relations of plants. II. The effect of temperature on increase in area of leaf surface and in dry weight of *Cucumis sativus*. Part I. The effect of temperature on the increase in area of leaf surface. *Ann. Bot.* **42**, 469–507.

Groom, P. (1908). Longitudinal symmetry in phanerogamia. *Phil. Trans. roy. Soc.* B **200**, 57–115.

Groom, P. (1909). The longitudinal symmetry of the Centrospermae. *Trans. Linn. Soc. Bot.* **7**, 267–302.

Gupta, B. (1961). Correlation of tissues in leaves. *Ann. Bot.* N.S. **25**, 65–71.

Guttenberg, H. von & Müller, H. (1957). Die laterale Anisophyllie von *Coleus hybridus* als Korrelationsphänomen. *Planta* **49**, 271–299.

Haber, A. H. (1962). Non-essentiality of concurrent cell division for degree of polarization of leaf growth. I. Studies with radiation-induced mitotic inhibition. *Am. J. Bot.* **49**, 583–589.

Haber, A. H. & Foard, D. E. (1963). Non-essentiality of concurrent cell divisions for degree of polarization of leaf growth, II. Evidence from untreated plants and from chemically induced changes of the degree of polarization. *Am. J. Bot.* **50**, 937–944.

Haccius, B. (1952). Über die Blattstellung einiger Hydrocharitaceen-Embryonen. *Planta* **40**, 333–345.

Harris, G. P. & Hart, E. M. H. (1964). Regeneration from leaf squares of *Peperomia sandersii* A.DC.: a relationship between rooting and budding. *Ann. Bot.* N.S. **28**, 509–526.

Harris, J. A., Sinnott E. W., Pennypacker, J. Y. & Durham, G. B. (1921). The vascular anatomy of dimerous and trimerous seedlings of *Phaseolus vulgaris*. *Am. J. Bot.* **8**, 63–102.

Harris, J. A., *et al.* (1921a). Correlations between anatomical characters in the seedling of *Phaseolus vulgaris*. *Am. J. Bot.* **8**, 339–365.

Harris, J. A., *et al.*, (1921b). The vascular anatomy of hemitrimerous seedlings of *Phaseolus vulgaris*. *Am. J. Bot.* **8**, 375–381.

Harris, J. A., *et al.* (1921c). The interrelationships of the number of the two types of vascular bundles in the transition zone of the axis of *Phaseolus vulgaris*. *Am. J. Bot.* **8**, 425–432.

Hayes, E. H. & Church, A. H. (1904). Mathematical notes on log. spiral systems and their application to phyllotaxis phenomena, pp. 327–347 in Church (1904).

Heathcote, D. G. & Idle, D. B. (1965). Nutation in seedling *Phaseolus multiflorus*. *Ann. Bot.* N.S. **29**, 563–577.

Hejnowicz, Z. (1959). Growth and cell division in the apical meristem of wheat roots. *Physiologia Pl.* **12**, 124–138.

Hendricks, H. V. (1919). Torsion studies in twining plants. *Bot. Gaz.* **68**, 425–440.

Hendricks, H. V. (1923). Torsion studies in twining plants. II. *Bot. Gaz.* **75**, 282–297.

Henry, A. (1910). On elm-seedlings showing Mendelian results. *J. Linn. Soc. Bot.* **39**, 290–300.

Hillson, C. J. (1963). Hybridization and floral vascularization. *Am. J. Bot.* **50**, 971–978.

Himmell, W. J. (1927). A contribution to the biophysics of *Podophyllum* petioles. *Bull. Torrey bot. Club* **54**, 419–451.

Hoffstadt, R. E. (1916). The vascular anatomy of *Piper methysticum*. *Bot. Gaz.* **62**, 115–132.

Holttum, R. E. (1955). Growth-habits of monocotyledons – variations on a theme. *Phytomorphology* **5**, 399–413.

Houghtaling, H. B. (1940). Stem morphogenesis in *Lycopersicum*: a quantitative study of cell size and number in the tomato. *Bull. Torrey bot. Club* **67**, 33–55.

Huber, B. (1935). *Aster linosyris*, ein neuer Typus der Kompasspflanzen (Gnomonpflanzen). *Flora, Jena* **129**, 113–119.

Isbell, C. L. (1931). Regenerative capacities of leaf and leaflet cuttings of tomato and of leaf and shoot cuttings of potato. *Bot. Gaz.* **92**, 192–201.

Jacobs, W. P. (1958). Further studies of the relation between auxin and abscission of *Coleus* leaves. *Am. J. Bot.* **45**, 673–675.

Jahn, E. (1941). Untersuchungen über die Zellzahl und Zellänge in der Epidermis der Internodien von *Vicia Faba Beih, bot. Zbl.* **60A**, 417–482.

Jensen, L. C. W. (1968). Primary stem vascular patterns in three subfamilies of the Crassulaceae. *Am. J. Bot.* **55**, 553–563.

Kaldewey, H. (1957). Wachstumsverlauf, Wuchsstoffbildung und Nutationsbewegungen von *Fritillaria meleagris* L. im Laufe der Vegetationsperiode. *Planta* **49**, 300–344.

Kaufman, P. B., Cassell, S. J. & Adams, P. A. (1965). On nature of intercalary growth and cellular differentiation in internodes of *Avena sativa*. *Bot. Gaz.* **126**, 1–13.

Keep, E. (1969). Accessory buds in the genus *Rubus* with particular reference to *R. idaeus* L. *Ann. Bot.* N.S. **33**, 191–204.

Kozlowski, T. T. & Clausen J. J. (1966). Shoot growth characteristics of heterophyllous woody plants. *Can. J. Bot.* **44**, 827–843.

Kröner, E. (1955). Experimentelle Beiträge zum Photoperiodismus der vegetativen Vermehrung der Gattung *Kalanchoe*. *Flora, Jena* **142**, 400–465.

Lems, K. (1964). Evolutionary studies in the Ericaceae. II. Leaf anatomy as a phylogenetic index in the Andromedeae. *Bot. Gaz.* **125**, 178–186.

Leshem, Y. & Koller, D. (1965). The control of runner development in the strawberry *Fragaria ananassa* Duch. *Ann. Bot.* N.S. **29**, 699–708.

Link, G. K. K. & Eggers, V. (1946). Mode, site, and time of initiation of hypocotyledonary bud primordia in *Linum usitatissimum* L. *Bot. Gaz.* **107**, 441–454.

Loeb, J. (1915). Rules and mechanism of inhibition and correlation in the regeneration of *Bryophyllum calycinum*. *Bot. Gaz.* **60**, 249–276.

Maini, J. S. (1966). Apical growth of *Populus* spp. I. Sequential pattern of internode, bud, and branch length of young individuals. *Can. J. Bot.* **44**, 615–622.

Maini, J. S. (1966a). Apical growth of *Populus* spp. II. Relative growth potential of apical and lateral buds. *Can. J. Bot.* **44**, 1581–1590.

Maksymowych, R. (1962). An analysis of leaf elongation in *Xanthium pennsylvanicum* presented in relative elemental growth rates. *Am. J. Bot.* **49**, 7–13.

Maksymowych, R. (1963). Cell division and cell elongation in leaf development of *Xanthium pennsylvanicum*. *Am. J. Bot.* **50**, 891–901.

Maksymowych, R. & Erickson, R. O. (1960). Development of the lamina in *Xanthium italicum* represented by the plastochron index. *Am. J. Bot.* **47**, 451–459.

Mann, L. K. (1960). Bulb organization in *Allium*: some species of the section *Molium*. *Am. J. Bot.* **47**, 765–771.

Martin, F. W. & Ortiz, S. (1963). Origin and anatomy of tubers of *Dioscorea floribunda* and *D. spiculiflora*. *Bot. Gaz.* **124**, 416–421.

McCully, M. E. & Dale, H. M. (1961). Variations in leaf number in *Hippuris*. A study of whorled phyllotaxis. *Can. J. Bot.* **39**, 611–625.

McLean Thompson, J. (1951). A further contribution to our knowledge of cauliflorous plants (with special reference to *Swartzia pinnata* Willd.). *Proc. Linn. Soc. Lond.* **162**, 212–222.

McLean Thompson, J. (1952). A further contribution to our knowledge of cauliflorous plants with special reference to the cannonball tree (*Couroupita guianensis* Aubl.). *Proc. Linn. Soc., Lond.* **163**, 233–250.

McVeigh, I. (1938). Regeneration in *Crassula multicava*. *Am. J. Bot.* **25**, 7–11.

Mehrlich, F. P. (1931). Factors affecting growth from the foliar meristems of *Bryophyllum calycinum*. *Bot. Gaz.* **92**, 113–140.

Namboodiri, K. K. & Beck, C. B. (1968). A comparative study of the primary vascular system of conifers. *Am. J. Bot.* **55**, 447–472.

Naylor, E. (1932). The morphology of regeneration in *Bryophyllum calycinum*. *Am. J. Bot.* **19**, 32–40.

Needham, J. (1942). *Biochemistry and morphogenesis*. (University Press, Cambridge.)

Nilson, E. B., Johnson, V. A. & Gardner, C. O. (1957). Parenchyma and epidermal cell length in relation to plant height and culm internode length in winter wheat. *Bot. Gaz.* **119**, 38–43.

Njoku, E. (1956). Studies in the morphogenesis of leaves. XI. The effect of light intensity on leaf shape in *Ipomoea caerulea*. *New Phytol.* **55**, 91–110.

Njoku, E. (1956a). The effect of defoliation on leaf shape in *Ipomoea caerulea*. *New Phytol.* **55**, 213–228.

Njoku, E. (1957). The effect of mineral nutrition and temperature on leaf shape in *Ipomoea caerulea*. *New Phytol.* **56**, 154–171.

Njoku, E. (1958). Effect of gibberellic acid on leaf form. *Nature, Lond.* **182**, 1097–1098.

O'Neill, T. B. (1961). Primary vascular organization of *Lupinus* shoot. *Bot. Gaz.* **123**, 1–9.

Ossenbeck, C. (1927). Kritische und experimentelle Untersuchungen an *Bryophyllum. Flora, Jena* **122**, 342–387.

Panje, R. R. (1961). On constriction bands and the system of integrated growth-zones in the leaf-blades of grasses. *Phytomorphology* **11**, 257–262.

Pearl, R. & Reed, L. J. (1920). On the rate of growth of the population of the United States since 1790 and its mathematical representation. *Proc. natn. Acad. Sci., U.S.A.* **6**, 275–288.

Philipson, W. R. & Balfour, E. E. (1963). Vascular patterns in dicotyledons. *Bot. Rev.* **29**, 382–404.

Philpott, J. (1953). A blade tissue study of leaves of forty-seven species of *Ficus. Bot. Gaz.* **115**, 15–35.

Plymale, E. L. & Wylie, R. B. (1944). The major veins of mesomorphic leaves. *Am. J. Bot.* **31**, 99–106.

Pratt, R. (1941). Validity of equations for relative growth constants when applied to sigmoid growth curves. *Bull. Torrey bot. Club.* **68**, 295–304.

Pray, T. R. (1962). Ontogeny of the closed dichotomous venation of *Regnellidium. Am. J. Bot.* **49**, 464–472.

Pray, T. R. (1963). Origin of vein endings in angiosperm leaves. *Phytomorphology* **13**, 60–81.

Rauh, W. (1937). Die Bildung von Hypocotyl- und Wurzelsprossen und ihre Bedeutung für die Wuchsformen der Pflanzen. *Nova Acta Leopoldina* N.F. **4**, 395.

Raunkiaer, C. (1919). Über Homodromie und Antidromie, insbesondere bei Gramineen. *Kgl. Danske Vidensk. Sellsk. Biol. Medd.* **1**, No. 12.

Reed, H. S. (1920). Slow and rapid growth. *Am. J. Bot.* **7**, 327–332.

Rees, A. R. (1964). The apical organization and phyllotaxis of the oil palm. *Ann. Bot.* N.S. **28**, 57–69.

Reynolds, M. E. (1942). Development of the node in *Ricinus communis*. *Bot. Gaz.* **104**, 167–170.

Richards, F. J. (1951). Phyllotaxis: its quantitative expression and relation to growth in the apex. *Phil. Trans. R. Soc.* B **235**, 509–564.

Rijven, A. H. G. C. (1968). Randomness in the genesis of phyllotaxis. I. The initiation of the first leaf in some Trifolieae. *New Phytol.* **67**, 247–256.

Rijven, A. H. G. C. (1969). Randomness in the genesis of phyllotaxis. II. Initiation of the third leaf in *Trigonella foenum-graecum* L. *New Phytol.* **68**, 377–386.

Robbins, W. J. (1957). Gibberellic acid and the reversal of adult *Hedera* to a juvenile state. *Am. J. Bot.* **44**, 743–746.

Robbins, W. J. (1960). Further observations on juvenile and adult *Hedera*. *Am. J. Bot.* **47**, 485–491.

Robertson, T. B. (1908). On the normal rate of growth of an individual, and its biochemical significance. *Arch. EntwMech. Org.* **25**, 581–614.

Rossetter, F. N. & Jacobs, W. P. (1953). Studies on abscission: the stimulating role of nearby leaves. *Am. J. Bot.* **40**, 276–280.

Sandt, W. (1925). Zur Kenntniss der Beiknospen. *Bot. Abh.* 7, 1–160.

Schanderl, H. (1929). *Sagittaria sagittifolia* als Kompasspflanze. *Planta* 7, 113–117.

Schanderl, H. (1932). Ökologische Untersuchungen an sogenannten Kompasspflanzen. *Planta* **16**, 709–762.

Schmucker, T. (1925). Rechts- und Linkstendenz bei Pflanzen. *Beih. bot. Zbl.* **41**, (1), 51–81.

Schmucker, T. (1933). Zur Entwicklungsphysiologie der schraubigen Blattstellung. *Planta* **19**, 139–153.

Schwartz, W. (1928). Zur Ätiologie der geaderten Panaschierung (1 Mitteilung). *Planta* 5, 660–680.

Schwartz, W. & Schwartz, H. (1928). Beiträge zur Biologie der plagiotropen Sprosse. I. Bildung oberirdischer Rhizome bei *Epilobium hirsutum* und *Lysimachia vulgaris*. *Flora, Jena* **123**, 21–29.

Shah, J. J. (1968). Axillary bud traces in certain dicotyledons. *Can. J. Bot.* **46**, 169–175.

Sinnott, E. W. (1914). The anatomy of the node as an aid in the classification of angiosperms. *Am. J. Bot.* **1**, 303–322.

Sinnott, E. W. (1930). The morphogenetic relationships between cell and organ in the petiole of *Acer*. *Bull. Torrey bot. Club* 57, 1–20.

Sinnott, E. W. & Durham, G. B. (1923). A quantitative study of anisophylly in *Acer*. *Am. J. Bot.* **10**, 278–287.

Sitte, P. (1957). Bildungsabweichungen an wirteligen Sprossen und ihre entwicklungsphysiologische und genetische Bedeutung. *Öst. bot. Z.* **104**, 234–302.

Skutch, A. F. (1927). Anatomy of leaf of banana, *Musa sapientum* L. var. Hort. Gros Michel. *Bot. Gaz.* **84**, 337–391.

Slade, B. F. (1957). Leaf development in relation to venation, as shown in *Cercis siliquastrum* L., *Prunus serrulata* Lindl. and *Acer pseudoplatanus* L. *New Phytol.* **56**, 281–300.

Smirnov, E. & Zhelochovtsev, A. N. (1931). Das Gesetz der Altersveränderungen der Blattform bei *Tropaeolum majus* L. unter Verschiedenen Beleuchtungs-bedingungen (Ein Beitrag sur Feldtheorie). *Planta* **15**, 299–354.

Smith, E. P. (1928). A comparative study of the stem structure of the genus *Clematis*, with special reference to anatomical changes induced by vegetative propagation. *Trans. R. Soc., Edinb.* **55**, 643–664.

Sparks, P. D. & Postlethwaite, S. N. (1967). Comparative morphogenesis of the dimorphic leaves of *Cyamopsis tetragonoloba*. *Am. J. Bot.* **54**, 281–285.

Sparks, P. D. & Postlethwaite, S. N. (1967a). Physiological control of the dimorphic leaves of *Cyamopsis tetragonoloba*. *Am. J. Bot.* **54**, 286–290.

Sparmann, G. (1961). Morphologische und biochemische Untersuchungen an vernalisierten und nicht vernalisierten Gerstenpflanzen. *Planta* **56**, 447–474.

Steffensen, D. M. (1968). A reconstruction of cell development in the shoot apex of maize. *Am. J. Bot.* **55**, 354–369.

Stein, O. L. & Steffensen, D. (1959). Radiation-induced genetic markers in the study of leaf growth in *Zea. Am. J. Bot.* **46**, 485–489.

Steward, F. C. (1954). Salt accumulation in plants: a reconsideration of the role of growth and metabolism. B. Salt accumulation in the plant body. *Symp. Soc. exp. Biol.* **8**, 393–406.

Stocker, O. (1926). Über transversale Kompasspflanzen. *Flora, Jena* **120**, 371–376.

Stocker, O. (1927). Über das Vorkommen von Kompasspflanzen. *Flora, Jena* **122**, 392.

Stoudt, H. N. (1938). Gemmipary in *Kalanchoë rotundifolia* and other Crassulaceae. *Am. J. Bot.* **25**, 106–110.

Stoutemeyer, V. T. & Britt, D. K. (1961). Effect of temperature and grafting on vegetative growth phases of Algerian ivy. *Nature, Lond.* **189**, 854.

Streitberg, H. (1954). Über die Heterophyllie bei Wasserpflanzen mit besonderer Berücksichtigung ihrer Bedeutung für die Systematik. *Flora, Jena* **141**, 567–597.

Sunderland, N. (1960). Cell division and expansion in the growth of the leaf. *J. exp. Bot.* **11**, 68–80.

Taylor, R. L. (1967). The foliar embryos of *Malaxis paludosa. Can. J. Bot.* **45**, 1553–1556.

Tenopyr, L. A. (1918). On the constancy of cell shape in leaves of varying shape. *Bull. Torrey bot. Club* **45**, 51–76.

Thompson, N. P. & Heimsch, C. (1964). Stem anatomy and aspects of development in tomato. *Am. J. Bot.* **51**, 7–19.

Thomson, B. F. & Miller, P. M. (1961). Growth patterns of pea seedlings in darkness and in red and white light. *Am. J. Bot.* **48**, 256–261.

Titman, P. W. & Wetmore, R. H. (1955). The growth of long and short shoots in *Cercidiphyllum. Am. J. Bot.* **42**, 364–372.

Uhrová, A. (1935). Über die Wechselbeziehungen der Knospen bei *Bryophyllum crenatum. Flora, Jena* **129**, 260–286.

Venning, F. D. (1949). Stimulation by wind motion of collenchyma formation in celery petioles. *Bot. Gaz.* **110**, 511–514.

Walker, W. S. (1960). The effects of mechanical stimulation and etiolation on the collenchyma of *Datura stramonium. Am. J. Bot.* **47**, 717–724.

Weber, H. (1950). Über das Wachstum des Rhizoms von *Butomus umbellatus* L. *Planta* **38**, 196–204.

Webster, B. D. & Steeves, T. A. (1958). Morphogenesis in *Pteridium aquilinum* (L.) Kuhn. General morphology and growth habit. *Phytomorphology* **8**, 30–41.

Whaley, W. G. & Whaley, C. Y. (1942). A developmental analysis of inherited leaf patterns in *Tropaeolum. Am. J. Bot.* **29**, 195–200.

White, D. J. B. (1954). The development of the runner-bean leaf with special reference to the relation between the sizes of the lamina and of the petiolar xylem. I. The relation between lamina area and petiolar xylem. *Ann. Bot.* N.S. **18**, 327–335.

White, D. J. B. (1954a). The development of the runner-bean leaf with special reference to the relation between the sizes of the lamina and of the petiolar xylem. II. The normal development of the bean leaf. *Ann. Bot.* N.S. **18**, 337–347.

White, D. J. B. (1955). The architecture of the stem apex and the origin and development of the axillary buds in seedlings of *Acer pseudoplatanus* L. *Ann. Bot.* N.S. **19**, 437–449.

White, D. J. B. (1957). Anisophylly of lateral shoots, *Ann. Bot.* N.S. **21**, 247–255.

White, R. A. (1968). A correlation between the apical cell and the heteroblastic leaf series in *Marsilea*. *Am. J. Bot.* **55**, 485–493.

Williams, W. T., Dore, J. & Patterson, D. G. (1957). Studies in the regeneration of horseradish. III. External factors. *Ann. Bot.* N.S. **21**, 627–632.

Wilson, C. L. (1924). Medullary bundle system in relation to primary vascular system in Chenopodiaceae and Amaranthaceae. *Bot. Gaz.* **78**, 175–199.

Winsor, C. P. (1932). The Gompertz curve as a growth curve. *Proc. natn. Acad. Sci., U.S.A.* **18**, 1–8.

Woltereck, I. (1928). Experimentelle Untersuchungen über die Blattbildung amphibischer Pflanzen. *Flora, Jena* **123**, 30–61.

Wylie, R. B. (1946). Relations between tissue organization and vascularization in leaves of certain tropical and subtropical dicotyledons. *Am. J. Bot.* **33**, 721–726.

Yarbrough, J. A. (1932). Anatomical and developmental studies of the foliar embryos of *Bryophyllum calycinum*. *Am. J. Bot.* **19**, 443–453.

Yarbrough, J. A. (1936). The foliar embryos of *Tolmiea menziesii*. *Am. J. Bot.* **23**, 16–20.

Yarbrough, J. A. (1936a). The foliar embryos of *Camptosorus rhizophyllus*. *Am. J. Bot.* **23**, 176–181.

Yarbrough, J. A. (1936b). Regeneration in the foliage leaf of *Sedum*. *Am. J. Bot.* **23**, 303–307.

Yin, H. C. (1941). Studies on the nyctinastic movement of the leaves of *Carica papaya*. *Am. J. Bot.* **28**, 250–261.

Zalenski, W. von (1902). Ueber die Ausbildung der Nervation bei verschiedenen Pflanzen. *Ber. dt. bot. Ges.* **20**, 433–440.

Zimmermann, M. H. & Tomlinson, P. B. (1968). Vascular construction and development in the aerial stem of *Prionium* (Juncaceae). *Am. J. Bot.* **55**, 1100–1109.

Index